The Open University

Mathematics Foundation Course Unit 30

GROUPS I

Prepared by the Mathematics Foundation Course Team

Correspondence Text 30

The Open University Press

Open University courses provide a method of study for independent learners through an integrated teaching system including textual material, radio and television programmes and short residential courses. This text is one of a series that make up the correspondence element of the Mathematics Foundation Course.

The Open University's courses represent a new system of university level education. Much of the teaching material is still in a developmental stage. Courses and course materials are, therefore, kept continually under revision. It is intended to issue regular up-dating notes as and when the need arises, and new editions will be brought out when necessary.

Further information on Open University courses may be obtained from The Admissions Office, The Open University, P.O. Box 48, Bletchley, Buckinghamshire.

The Open University Press
Walton Hall, Bletchley, Bucks

First Published 1971

Printed in Great Britain by
J W Arrowsmith Ltd, Bristol 3

SBN 335 01029 6

Contents

Objectives

The principal objective of this unit is to introduce the concept of a *group*.

After working through this unit you should be able to:

(i) explain what is meant by a symmetry operation;
(ii) construct the symmetry table for a given (simple) geometric figure;
(iii) state a set of axioms for a group and test particular examples for group structure;
(iv) make simple deductions from group axioms;
(v) find symmetry groups of algebraic expressions;
(vi) define the term *permutation*;
(vii) state Cayley's theorem and understand its proof.

Note

Before working through this correspondence text, make sure you have read the general introduction to the mathematics course in the Study Guide, as this explains the philosophy underlying the whole course. You should also be familiar with the section which explains how a text is constructed and the meanings attached to the stars and other symbols in the margin, as this will help you to find your way through the text.

Structural Diagram

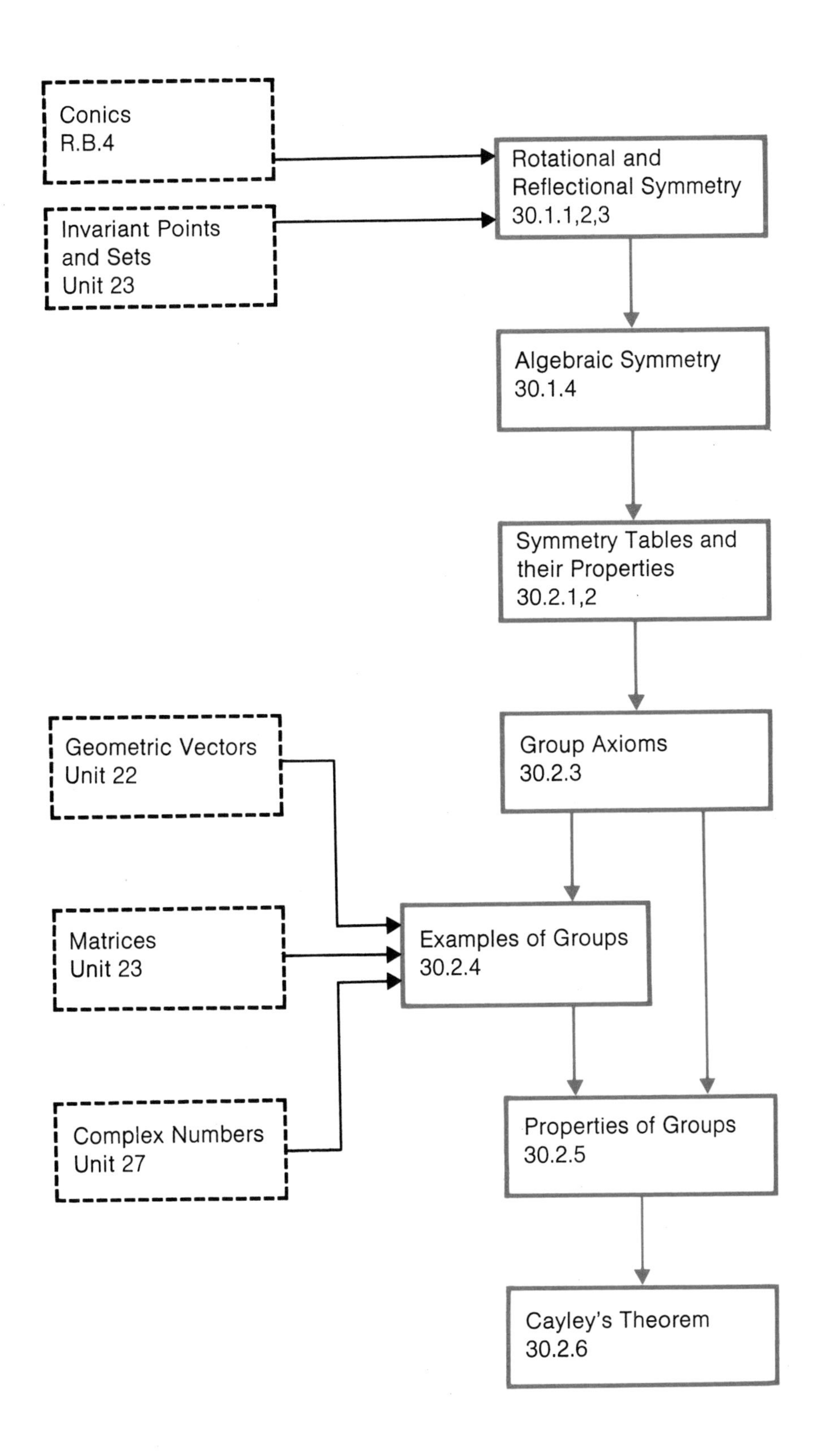

Glossary

Page

Terms which are defined in this glossary are printed in CAPITALS.

Term	Definition	Page
ABELIAN GROUP	An ABELIAN GROUP is a GROUP $(G, \circ)$ such that $$g \circ h = h \circ g$$ for all $g, h \in G$.	41
CYCLIC GROUP	A CYCLIC GROUP is a GROUP GENERATED by a single element.	33
EQUALITY OF SYMMETRY OPERATIONS	Two SYMMETRY OPERATIONS S and T on a figure (or algebraic expression) are EQUAL if and only if the image of each point of the figure (or the image of the expression) is the same under S as under T.	21
GENERATE	A subset S of a GROUP $(G, \circ)$ GENERATES G if for every $a \in G$ we can express a as a combination (using $\circ$) of the elements of S.	26
GROUP	A GROUP $(G, \circ)$ is a set G with a binary operation $\circ$ defined on it, which satisfies four particular conditions. See section 30.2.3.	27
ORDER	A GROUP $(G, \circ)$ comprising n elements is said to have ORDER n. If G is not a finite set, the group $(G, \circ)$ is said to have INFINITE ORDER.	27
PERMUTATION	A PERMUTATION on a set of elements S is a one-one function f such that $f(S) = S$.	26
SUBGROUP	$(S, \circ)$ is a SUBGROUP of a GROUP $(G, \circ)$ if S is a subset of G such that $(S, \circ)$ is itself a group.	37
SYMMETRY OPERATION	A SYMMETRY OPERATION on a figure in space is a rigid mapping of the space to itself, under which the figure is invariant.	1
	A SYMMETRY OPERATION on an algebraic expression is a mapping under which the expression is invariant.	19

Notation

The symbols are presented in the order in which they appear in the text.

		Page
$\circ$	A binary operation.	12
$(G, \circ)$	A group consisting of the set G with the binary operation $\circ$ defined on it.	27
e	The identity element in a group.	27
R^+	The set of positive real numbers.	29
R^-	The set of negative real numbers.	29
$\exists_x$	The existential quantifier ("there exists an x such that").	38
$\forall_x$	The universal quantifier ("for all x").	38
g^{-1}	The inverse of the element g of a group $(G, \circ)$.	39
p_h	The permutation defined by the element h of a group $(G, \circ)$: $p_h : g \longmapsto h \circ g \quad (g \in G)$.	43

Bibliography

H. Liebeck, *Algebra for Scientists and Engineers* (John Wiley, 1969).

The approach to group theory adopted in this book is similar to ours. Symmetry operations (transformations) are discussed in Chapter 6, and the concept of a group is introduced in Chapter 7.

F. M. Hall, *An Introduction to Abstract Algebra*, Vol. I (Cambridge University Press, 1966).

This book gives a clear account of group theory and discusses groups of symmetry operations and permutations. The relevant chapters are Chapters 11, 12 and 13.

Acknowledgements

Grateful acknowledgement is made to the following sources for illustrations used in this correspondence text:
British Film Institute, decorative curtain, p. 18; Camera Press, Pisa, p. 13; French Government Tourist Office, Chartres window, p. 13; Mansell Collection, Greek vases, pp. 3 and 17, portraits of Plato, p. 16 and Cayley, p. 47; Paul Popper Limited, morning-glory, p. 33; Royal Institute of British Architects, dome, p. 2; Royal Meteorological Society, snow-flakes, p. 2 and cover; Science Museum, London, portraits of Whitehead, p. 1, Galois, p. 19, Klein, p. 36, Abel, p. 41.

30.0 INTRODUCTION

30.0

Introduction
* *

In this unit we use the idea of *symmetry* to introduce the mathematical concept of a *group*. A *group* consists of a set with a binary operation defined on it, which satisfies certain conditions.

The philosopher A. N. Whitehead (1861–1947) observed that "To see what is general in what is particular and what is permanent in what is transitory is the aim of scientific thought". This remark is particularly relevant to group theory.

A. N. Whitehead

For instance, in the television programme associated with this unit we see the same group arising out of three apparently different situations; by constructing a group to represent each situation we extract the *general* structure from the *particular* details, and by this process we see that details sometimes obscure more general (and perhaps more fundamental) properties.

And what of the second part of Whitehead's remark? Symmetry is concerned with certain mappings, called *symmetry operations*, which leave sets of points unchanged. (In this unit we shall use the word *mapping* rather than *function*, because it is natural to speak of mapping one set to another. However, all the mappings which we discuss are in fact one-one functions.) In recent units, we have discussed a number of mappings of the plane to itself. If we draw a figure in the plane, then some such mappings will leave the shape and size of the figure unaltered, whatever the figure may be. For example, the shape and size of any figure are unaltered by a *rotation* of the plane about the origin, although, in general, the *position* of the figure is altered. What happens to a circle, with its centre at the origin, if we rotate the plane through any angle about the origin? It stays where it is: its *position* is unchanged. What happens to a square, with its centre at the origin, if we rotate the plane through a right-angle about the origin? It stays where it is: its *position* is unchanged.

We have already discussed elements in the domain of a function which are unchanged by the function (see *Unit 23*, section 23.1.1). We defined an *invariant element* under a function f to be an element a of the domain of f such that

$$f(a) = a,$$

and an *invariant set* under a function f to be a subset A of the domain of f such that

$$f(A) = A.$$

Investigation of invariant properties of particular mappings can give considerable insight into a situation. For example, how might we describe the idea of symmetry? The circle has symmetry and we can recognize this mathematically by the fact that *any* rotation in the plane about its centre leaves the circle where it is: the circle is an invariant set under the rotation mapping. A square has symmetry, but not the same symmetry as the circle. Only certain rotations $\left(\text{i.e. by integer multiples of } \frac{\pi}{2}\right)$ map the square to itself. Do these rotations completely describe the symmetry of the square? Are there any other mappings under which the square is an invariant set? The answers to questions of this sort enable us to build up an idea of the properties of the square. The same principle can be adopted in new and unfamiliar situations. Finding out which mappings leave certain properties unchanged sometimes helps the investigation of a new structure.

In this unit we are concerned with the symmetry of patterns which surround us.

From crystals and molecular structure to massive architecture, symmetry plays an important role.

In the course of the unit we shall try to formalize what we mean by symmetry (in terms of mappings) so that we can analyse the symmetry of a given figure mathematically.

Along the way we shall spend a little time developing a useful tool for sketching graphs, making use of any symmetry which can be identified algebraically before sketching begins.

Once we have discussed the symmetry of a few geometric figures, we shall observe certain properties which are shared by many other mathematical structures, leading us to define the abstract concept of a *group*.

But that is not all there is to groups. Group theory is itself a whole branch of mathematics, and it all stems from the four defining properties which we shall establish in this unit. But even for those who are not interested in group theory for its own sake, it is a most important mathematical idea. It underlies so much of mathematics that it is an important tool in mathematical and scientific research.

30.1 SYMMETRY

30.1

30.1.0 Introduction

30.1.0

Introduction
* *

The word *symmetry* generally has two slightly different connotations. The first conveys a sense of harmony and proportion, and is the basis of some rational analyses of aesthetic experience. The word itself is of Greek origin, and the concept has always been of major importance in art and architecture. The pyramids of Egypt and the ceremonial centres of Central America, Chinese and Indian temples, "primitive" decorations on pottery and cloth found all over the world, all demonstrate man's apparent desire for the orderliness which is implied by the regularity and symmetry of his design.

The second connotation of symmetry is the precise, geometric symmetry best exemplified by the bilateral symmetry of left and right so often found in nature. An object is said to be *symmetric about a plane* if the object is carried into itself by reflection in the plane, where we use the word *reflection* in the sense of a reflection in a mirror. Notice that *is carried into itself by reflection* contains the notion of a mapping, and *into itself* contains the notion of invariance of the object under the mapping. In other words, an object is symmetric about a plane if it is mapped to itself by the reflection of each point of the object in the given plane. We shall refer to the plane as a plane of symmetry of the object. In the case of a planar object we shall refer to a line of symmetry (rather than to a plane perpendicular to the object through a particular line). This is the starting point for a precise mathematical analysis of symmetry.

The two ideas of symmetry are not far apart. For instance, although most people seem to be approximately symmetric about a vertical plane, there are small details which destroy the apparent precise geometrical symmetry. Few people have arms which are exactly the same length, or hands which are exactly the same size. A similar situation holds for most objects which are illustrated in this text. At first sight they appear symmetric, but they deviate from symmetry in certain details. In fact it is frequently the case that a shape or collection of shapes is appealing because of its slight variation from being symmetric. Nevertheless, it is from observations of natural objects, from molecules to planets, that we arrive at our notion of geometric symmetry.

In section 30.1.1 we shall examine reflectional symmetry and its application to sketching graphs. In section 30.1.2 we shall take up the idea of rotational symmetry, and in section 30.1.3 we shall give examples of the combinations of the two that occur in nature and geometry, and we shall briefly mention translational symmetry.

30.1.1 Reflectional Symmetry

30.1.1

Discussion
* *

Reflectional symmetry is found in the pictorial graphs of some mappings, and can frequently be spotted algebraically *before* the graph is drawn, thus saving a lot of work.

Example 1

Example 1

The graph of the function

$$x \longmapsto x^2 \qquad (x \in R)$$

is symmetric about the y-axis. That is, the graph has the same shape on each side of the y-axis. Alternatively, we could say that the graph has the same height at equal distances to the left and right of the y-axis, or, more precisely, that if the point (x_0, y_0) lies on the graph, then so also does the point $(-x_0, y_0)$.

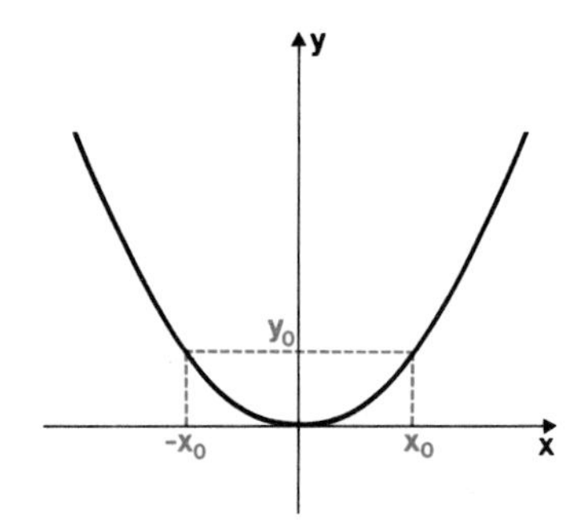

We shall refer to the equation

$$y = x^2$$

as the *equation of the graph*. (Of course, any multiple of this equation determines the same graph.) ■

Main Text

In the introduction to this section, we noticed that the ideas of symmetry and mapping are closely related.

If we map each point of the plane (x, y) to the point $(-x, y)$, then we are simply performing a reflection in the y-axis. For example,

$$(-3, 2) \longmapsto (3, 2)$$
$$(3, 2) \longmapsto (-3, 2)$$
$$(3, -4) \longmapsto (-3, -4)$$
$$(-3, -4) \longmapsto (3, -4)$$

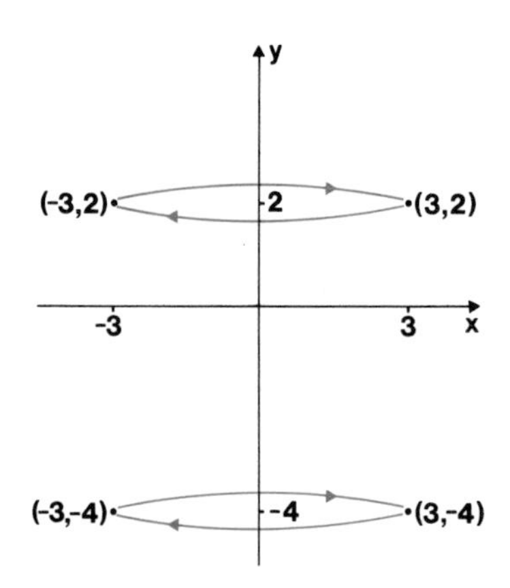

Our criterion for a graph to be symmetric about the y-axis can be interpreted in terms of this reflection mapping: under the mapping

$$(x, y) \longmapsto (-x, y) \qquad ((x, y) \in R^2)$$

the set of points which make up the graph is mapped to itself; that is, the set of points is an invariant set under the reflection mapping. We can test for this symmetry by looking at the equation of the graph, changing x to $-x$ and seeing whether the equation is unchanged. For example, the equation

$$y = x^2$$

becomes

$$y = (-x)^2$$

i.e.

$$y = x^2;$$

the equation is unchanged when x is replaced by $-x$. The equation

$$y = x^3 - x$$

becomes

$$y = (-x)^3 - (-x)$$
$$= -x^3 + x,$$

and this equation is not the same as the original equation.

The equation $y = x^2$ is the equation of a graph which is symmetric about the y-axis: the graph with equation $y = x^3 - x$ is not symmetric about the y-axis.

So far we have achieved two things. We can tell if the graph of an equation is symmetric about the y-axis, and, given the equation of a graph, we can find the equation of the reflection of the graph in the y-axis. For instance, we found the equation of the reflection of the graph of

$$y = x^3 - x$$

to be

$$y = -x^3 + x.$$

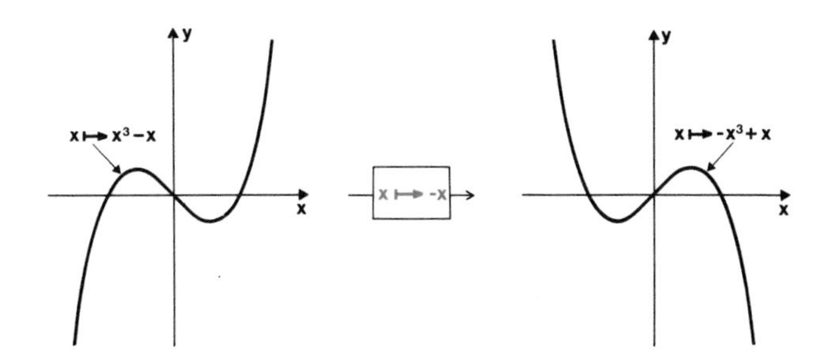

In the examples we have given so far, the graphs have been rather easy to draw and we knew all about their symmetry without any algebraic manipulation. It is interesting to see that this symmetry can be expressed algebraically. There is also a practical motive: we may be able to see that the graph of a new or difficult equation is symmetric without drawing it first.

We are extending our ideas of symmetry to algebraic expressions.

Exercise 1

Exercise 1
(2 minutes)

For each of the following equations, state whether or not the corresponding graph is symmetric about the y-axis. In each case $x \in R$.

(i) $y = \cos x$
(ii) $y = \sin x$
(iii) $x^2 + y^2 = 1$
(iv) $x^2 + xy + y^2 = 1$
(v) $y = (x + 1)(x - 1)$
(vi) $|x| + |y| = 1$
(vii) $y = \cos x + \sin x$
(viii) $y = \sin^2 x$. ■

Exercise 2

Exercise 2
(4 minutes)

What mapping will reflect the points of the plane in the line with equation $x = a$? Using your mapping, check whether or not the graph of each of the equations

$$y = (x - a)^2, \qquad y^2 + (x - a) = 1$$

is symmetric about the line with equation $x = a$. ■

Solution 1

Solution 1

The graphs corresponding to the following equations *are* symmetric about the y-axis: (i), (iii), (v), (vi) and (viii). ■

Solution 2

Solution 2

If $x > a$, the x-co-ordinate must be *decreased* by $2(x - a)$. If $x < a$, the x-co-ordinate must be *increased* by $2(a - x)$. In either case we can say that the x-co-ordinate must be *increased* by $2(a - x)$. The mapping is

$$(x, y) \longmapsto (x + 2(a - x), y)$$

i.e.

$$(x, y) \longmapsto (2a - x, y) \qquad ((x, y) \in R^2).$$

Another way of looking at it is as follows. In the first place map the whole plane by mapping

$$(x, y) \longmapsto (x - a, y),$$

which maps the line with equation $x = a$ to the y-axis. Then to reflect in the y-axis, we have to map

$$(x, y) \longmapsto (-x, y).$$

We now have to put everything back where it belongs, so finally we map

$$(x, y) \longmapsto (x + a, y).$$

Combining these mappings we have

$$(x, y) \longmapsto (x - a, y) \longmapsto (a - x, y) \longmapsto (2a - x, y).$$

The graph with equation $y = (x - a)^2$ is symmetric about the line with equation $x = a$. The graph with equation $y^2 + (x - a) = 1$ is not.

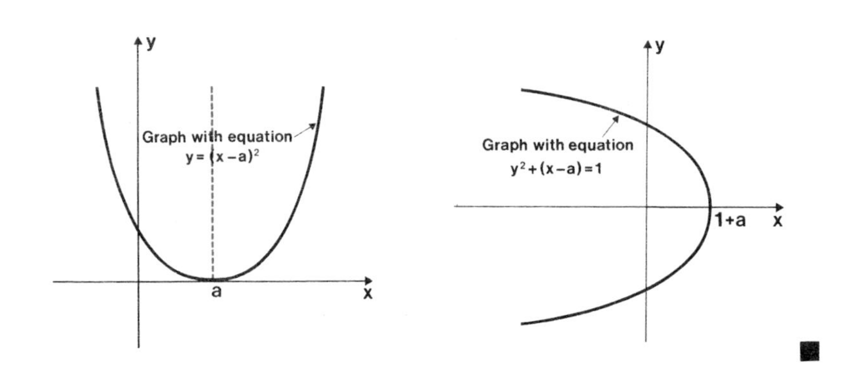

■

So far we have discussed only symmetry about the y-axis or lines parallel to it. We can deal with symmetry about the x-axis in an analogous fashion, by observing that each point (x, y) will be mapped to the point $(x, -y)$ by a reflection in the x-axis. Under the mapping

Discussion
* *

$$(x, y) \longmapsto (x, -y) \qquad ((x, y) \in R^2),$$

we find, for example, that the equation

$$y = x^2$$

becomes

$$-y = x^2,$$

so the graph of $y = x^2$ is not symmetric about the x-axis; and that the equation

$$y^2 - x^2 = 0$$

becomes

$$y^2 - x^2 = 0,$$

so the graph of $y^2 - x^2 = 0$ is symmetric about the x-axis.

Exercise 3

Exercise 3
(2 minutes)

Which of the equations in Exercise 1 correspond to graphs which are symmetric about the x-axis? ■

Exercise 4

Exercise 4
(3 minutes)

Which mapping will reflect the points of the plane in the line with equation $y = b$? ■

Discussion
* *

Another very common symmetry operation is a reflection in the line with equation $y = x$. Under such a reflection, the points on the x- and y-axes interchange positions, as shown in the next diagram. The reflection mapping is

$$(x, y) \longmapsto (y, x) \qquad ((x, y) \in R^2).$$

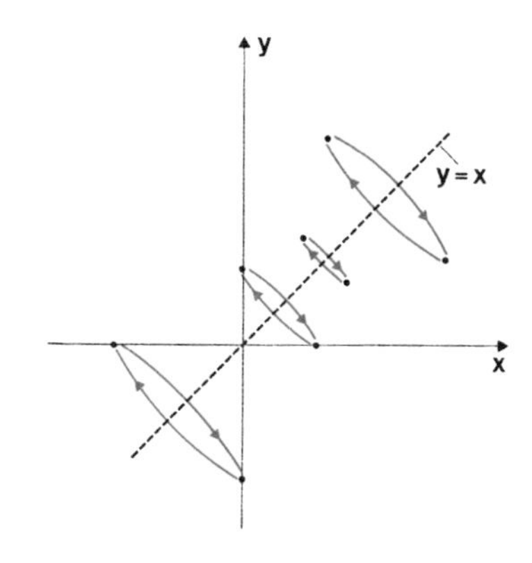

Thus

$$x^2 + xy + y^2 = 3$$

becomes

$$y^2 + yx + x^2 = 3;$$

hence the corresponding graph is symmetric about the line $y = x$.

But

$$y = x^2$$

becomes

$$x = y^2$$

and so the graph with equation $y = x^2$ is not symmetric about the line $y = x$.

Exercise 5

Exercise 5
(2 minutes)

Which of the equations in Exercise 1 has a graph which is symmetric about the line with equation $y = x$? ■

Exercise 6

Exercise 6
(3 minutes)

Which mapping has the effect of reflecting the plane in the line with equation $y = -x$? ■

(*continued on page 8*)

Solution 3

Solution 3

Equations (iii) and (vi) correspond to graphs which are symmetrical about the x-axis. ■

Solution 4

Solution 4

Reasoning as in Exercise 2 gives the mapping

$$(x, y) \longmapsto (x, 2b - y) \qquad ((x, y) \in R^2).$$ ■

Solution 5

Solution 5

Cases (iii), (iv), and (vi) are symmetric about the line with equation $y = x$. ■

Solution 6

Solution 6

$$(x, y) \longmapsto (-y, -x) \qquad ((x, y) \in R^2).$$ ■

(*continued from page 7*)

Exercise 7

Exercise 7
(4 minutes)

Which points of the plane are invariant under the following mappings? (Note: we are asking for individual invariant points as opposed to sets of invariant points, that is points (a, b) which map to (a, b). A set of points can be invariant even though the individual points of the set are not.)

(i) $(x, y) \longmapsto (-x, y)$
(ii) $(x, y) \longmapsto (x, -y)$
(iii) $(x, y) \longmapsto (y, x)$
(iv) $(x, y) \longmapsto ((A + 1)x + By + C, Ax + (B + 1)y + C)$ where A, B, C are real numbers. ■

Main Text
* * *

Symmetry about the line with equation $y = x$ arises in sketching the graphs of inverse functions. If f is a one-one function, $f: A \longmapsto B$, where $A, B \subseteq R$, then f has an inverse function $g: B \longmapsto A$. How are the graphs of f and g related? If we notice that the equations $y = f(x)$ and $x = g(y)$ say exactly the same thing, and hence have the same graphs, then obviously the graph with equation $y = g(x)$ is the same as the graph of the equation obtained by changing x to y and y to x in the equation $y = f(x)$. This is equivalent to reflecting the graph of f in the line with equation $y = x$.

In the following diagrams we show:

(i) the graph of the function

$$x \longmapsto \ln x \qquad (x \in R^+)$$

reflected in the line $y = x$ to give the graph of the function

$$x \longmapsto \exp x \qquad (x \in R)$$

(and vice versa);

(ii) the graph of the function

$$x \longmapsto 3x + 5 \qquad (x \in R)$$

reflected in the line $y = x$ to give the graph of the function

$$x \longmapsto \tfrac{1}{3}(x - 5) \qquad (x \in R)$$

(and vice versa).

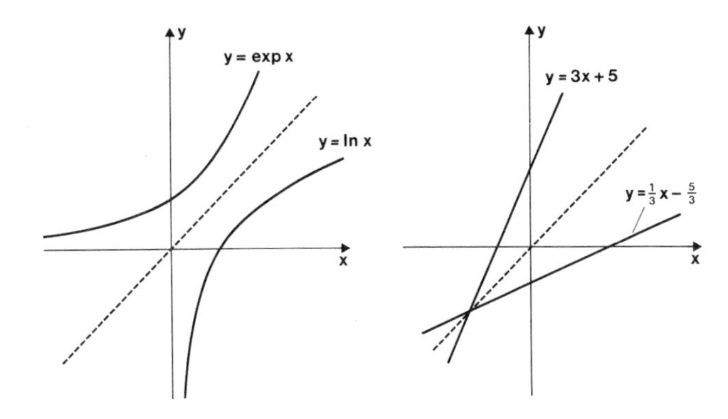

Similar remarks apply to mappings which are not functions. You may recall that in *Unit 1, Functions* we defined the graph of a mapping to be a set of ordered pairs. The reverse mapping is obtained by reversing the order of the pairs and interchanging the domain and the set of images. For a mapping whose domain and codomain are subsets of R, this corresponds to reflecting the pictorial graph in the line with equation $y = x$.

It was mentioned at the beginning of this section that recognizing symmetry in an equation can save a lot of work in sketching the graph. The next example is an extreme case of this.

Example 2

Example 2

Plot the graph with equation

$$|x| + |y| + |x + y| + |x - y| = 3.$$

At first sight this looks difficult. But can we detect any symmetry?

(i) Performing the mapping

$$(x, y) \longmapsto (-x, y) \qquad ((x, y) \in R^2),$$

which reflects the graph in the y-axis, gives the equation

$$|-x| + |y| + |-x + y| + |-x - y| = 3.$$

This is just the original equation back again, because

$$|-x| = |x| \text{ and } |-x + y| = |x - y| \text{ and } |-x - y| = |x + y|.$$

Therefore the graph must be symmetric about the y-axis, and so we need only plot points to the right of and on the y-axis, since we can obtain the others by reflection.

(ii) Performing the mapping

$$(x, y) \longmapsto (x, -y) \qquad ((x, y) \in R^2),$$

which reflects the graph in the x-axis, gives the equation

$$|x| + |-y| + |x - y| + |x + y| = 3,$$

which is the original equation again. Therefore the graph is symmetric about the x-axis as well, and so we need only plot points which lie to the right of or on the y-axis *and* above or on the x-axis, since reflection in the x-axis yields all the points to the right of the y-axis, and then, as we have seen, we get all the remaining points by reflection in the y-axis.

(iii) Performing the mapping

$$(x, y) \longmapsto (y, x) \qquad ((x, y) \in R^2),$$

which reflects the graph in the line $y = x$, also leaves the equation unchanged. Thus the graph is also symmetric about the line $y = x$,

(*continued on page 10*)

Solution 7

Solution 7

(i) If (a, b) is an invariant element under the mapping, then

$$(a, b) = (-a, b)$$

which implies that $a = 0$. Thus the points on the y-axis are invariant under the mapping.

(ii) Points on the x-axis.

(iii) For (a, b) to be invariant,

$$(a, b) = (b, a).$$

Points on the line with equation $x = y$ are invariant.

(iv) For (a, b) to be invariant

$$a = (A + 1)a + Bb + C$$

$$b = Aa + (B + 1)b + C.$$

Either equation is equivalent to

$$Aa + Bb + C = 0$$

and so points on the line with equation $Ax + By + C = 0$ are invariant. ■

(*continued from page 9*)

and so we only need to plot points which lie on or below the line $y = x$ *and* on or above the x-axis, and then by successive reflections, plot all the remaining points.

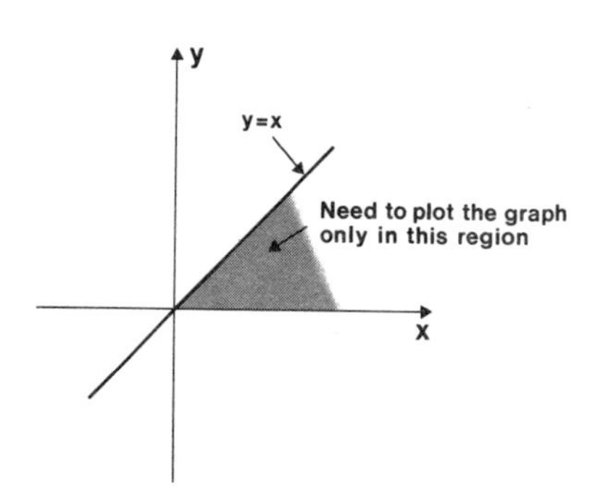

Now we are almost ready to start plotting a few points! If a point with co-ordinates (x, y) lies in the specified "plotting region", then its co-ordinates must satisfy

$$y \geqslant 0 \qquad ((x, y) \text{ is on or above the } x\text{-axis})$$

$$y \leqslant x \qquad ((x, y) \text{ is on or below the line } y = x)$$

i.e.

$$0 \leqslant y \leqslant x.$$

Now under these conditions we know that

$$|x| = x \text{ since } x \geqslant 0;$$

$$|y| = y \text{ since } y \geqslant 0;$$

$$|x + y| = x + y \text{ since } x \geqslant 0, y \geqslant 0;$$

$$|x - y| = x - y \text{ since } x \geqslant y.$$

Therefore in the "plotting region" the equation reduces to

$$x + y + (x + y) + (x - y) = 3$$

which is just

$$3x + y = 3.$$

The successive stages of drawing the graph are:

(i) Draw the part of the graph with equation $3x + y = 3$ which is in the "plotting region":

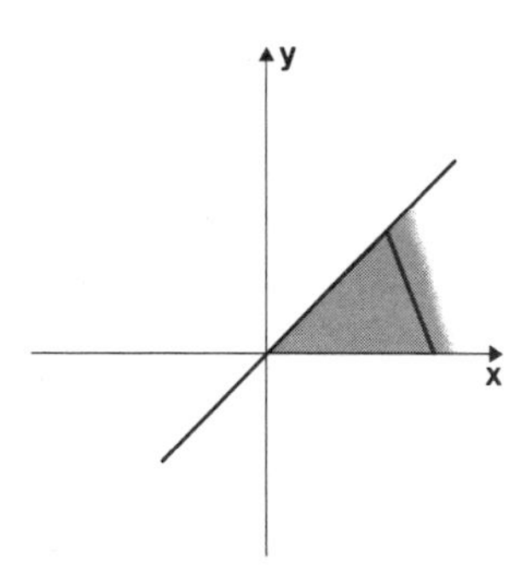

(ii) reflect this graph in the line $y = x$:

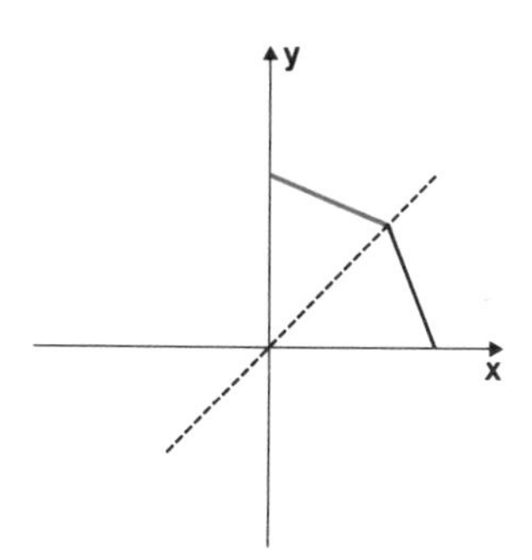

(iii) reflect this graph in the y-axis:

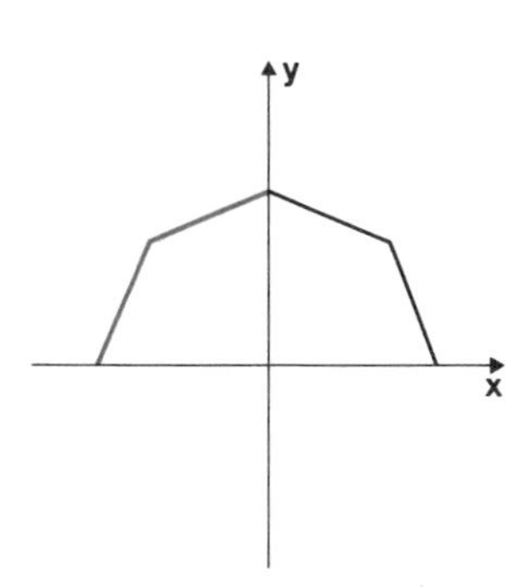

(iv) reflect this graph in the x-axis to obtain the graph with equation

$$|x| + |y| + |x + y| + |x - y| = 3.$$

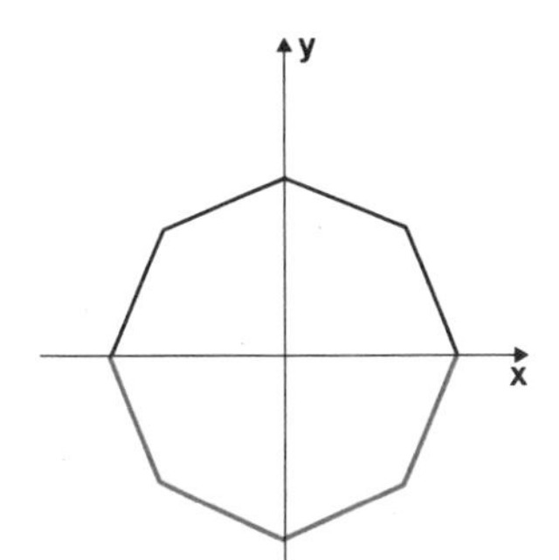

■

Exercise 8

Exercise 8
(5 minutes)

(i) If A is an invariant set under mappings f and g; that is,

$$f(A) = A, \qquad g(A) = A,$$

show that A is an invariant under $f \circ g$ and $g \circ f$. (Assume that f and g are suitably defined for $f \circ g$ and $g \circ f$ to exist.)

(ii) Assume that you have discovered that the graph of some equation is symmetric about the y-axis and about the line $y = x$. Does it follow that the graph is symmetric about the line $y = -x$? ■

30.1.2 Rotational Symmetry

30.1.2

Discussion
★ ★

Reflections are not the only possible symmetry operations. In the Introduction to this text we remarked that some sets are invariant under rotations. Consider the following examples.

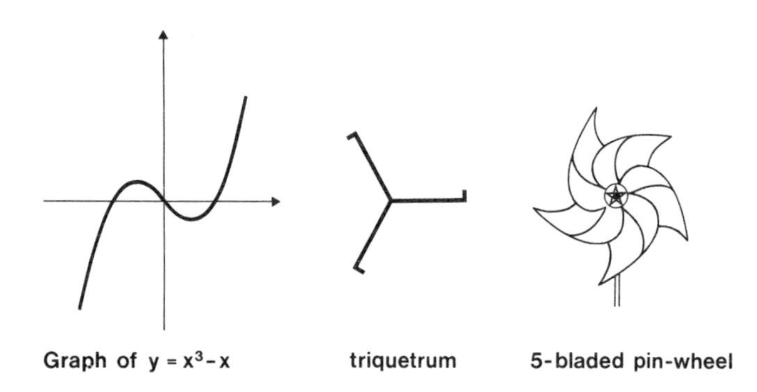

Graph of $y = x^3 - x$ triquetrum 5-bladed pin-wheel

Each of these three objects has rotational symmetry.

In each case, think of a pin stuck in the "centre" of the figure, and rotate the figure about the pin as axis. Again we are performing a mapping of the plane to itself, but this time each point of the plane is rotated about the "centre" through some angle θ. The first thing to notice is that if $\theta = 2\pi$, then every point is an invariant point. In other words, rotation through 2π has the same effect on each point of the plane as rotation through an angle 0. Obviously then we do not distinguish between these two rotations.

Now the first figure is also invariant under a rotation through an angle π, and the two symmetry operations, rotations through 0 and π, are the only two distinct rotational symmetry operations of the graph. The rotational axis in this case is called a *two-fold axis of symmetry.*

In the case of the triquetrum, the distinct rotations under which the figure is invariant are rotations through $0, \frac{2\pi}{3}$ and $\frac{4\pi}{3}$; this figure has a *three-fold axis of symmetry.*

The pin-wheel is invariant under the rotations through $0, \frac{2\pi}{5}, \frac{4\pi}{5}, \frac{6\pi}{5}$, and $\frac{8\pi}{5}$; it has a *five-fold axis of symmetry.*

Of course, a circle is invariant under any rotation whatsoever about its centre. All the previous examples admit only a finite number of distinct symmetry operations, but the circle admits infinitely many; it has an *infinite axis of symmetry*. Any theory of symmetry which we develop will have to cover this situation as well.

Rotational symmetry is very common in ancient pottery designs and in some famous architectural instances, particularly the buildings at Pisa, and also in the design of Gothic rose-windows.

Pisa

Chartres window

Solution 30.1.1.8

Solution 30.1.1.8

(i) We have

$$(f \circ g)(A) = f(g(A))$$
$$= f(A) \qquad (A \text{ is invariant under } g)$$
$$= A \qquad (A \text{ is invariant under } f).$$

Similarly,

$$(g \circ f)(A) = A.$$

(ii) Such a graph is invariant under each of the mappings

$$f:(x, y) \longmapsto (y, x) \qquad \text{(reflect in line with equation } y = x)$$
$$g:(x, y) \longmapsto (-x, y) \qquad \text{(reflect in } y\text{-axis)},$$

and we are asked whether it follows that the graph is invariant under the mapping

$$h:(x, y) \longmapsto (-y, -x).$$

We have

$$f \circ g:(x, y) \longmapsto (y, -x),$$
$$g \circ f:(x, y) \longmapsto (-y, x).$$

Neither of these is h, but

$$g \circ (f \circ g):(x, y) \longmapsto (-y, -x)$$

is h. So, by (i), the set is also invariant under h.

We can also deduce this result pictorially:

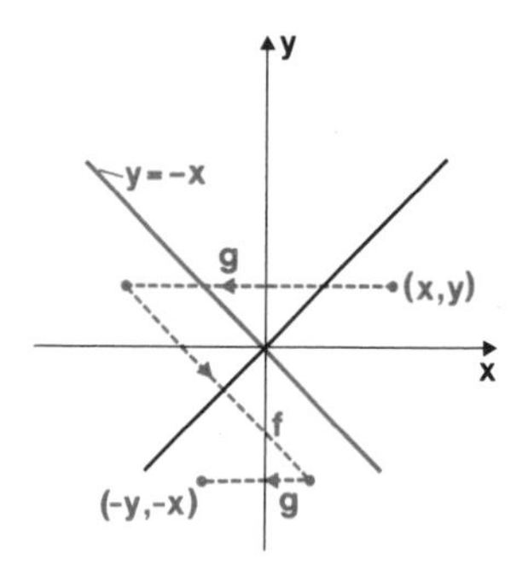

A reflection in the y-axis, followed by a reflection in the line $y = x$, followed by another reflection in the y-axis, is the same as a reflection in the line $y = -x$. ■

30.1.3 Combined Symmetries

30.1.3

Discussion
* *

In the examples considered so far we have concentrated on two kinds of symmetry (rotational and reflectional) and the examples have been essentially two-dimensional in nature. In this section we shall give some examples (not necessarily two-dimensional) of geometrical and natural objects with a high degree of geometric symmetry, usually of more than one type. We begin by looking at the so-called Platonic solids.

The Platonic solids are defined by the fact that in each case:

all edges have the same length;
all faces are congruent (i.e. have the same shape and size);
any two vertices are indistinguishable;
no two edges have a point in common other than a vertex.

They are:

(i) the regular tetrahedron

(ii) the cube

(iii) the octahedron

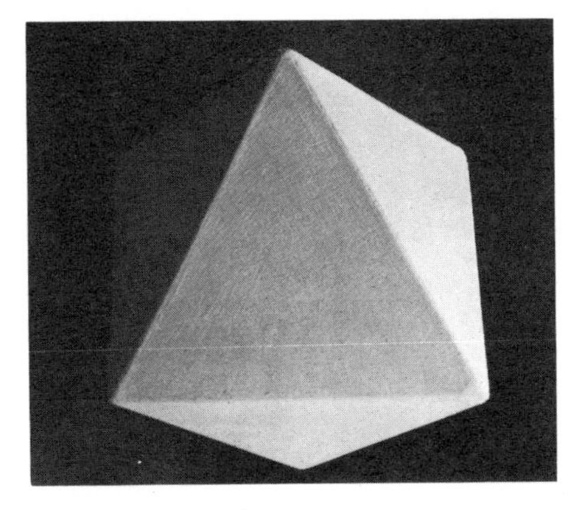

(iv) the icosahedron (the faces of which are 20 regular triangles)

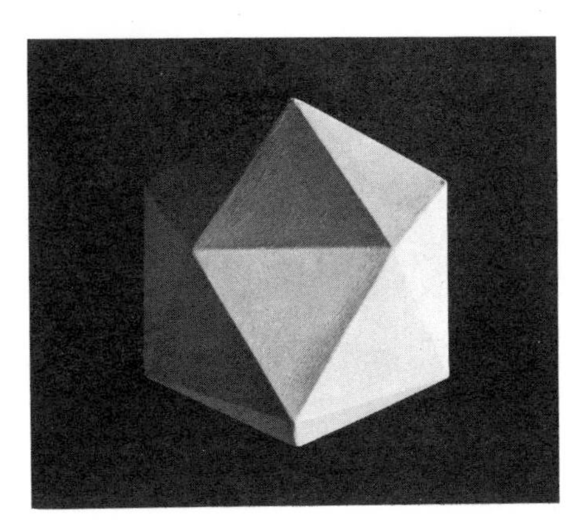

(v) the pentagondodecahedron (the faces of which are 12 regular pentagons)

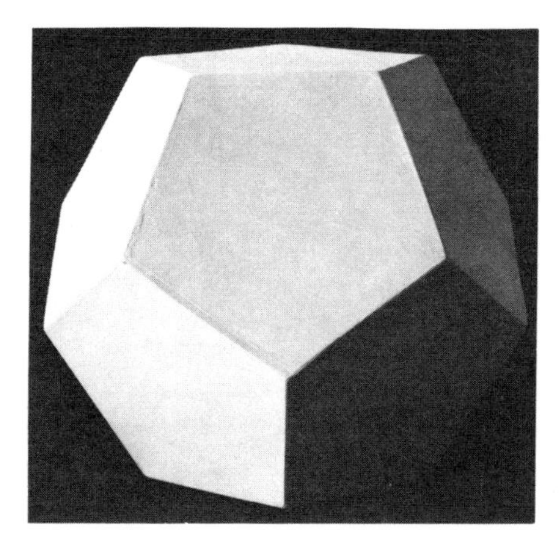

They are the only regular polyhedra in three-dimensional space.

The Platonic solids are so-named, not because Plato (428–348 B.C.) discovered them, but because he associated the regular tetrahedron, cube, octahedron and icosahedron with the four elements of fire, earth, air and water respectively; he associated the pentagondodecahedron with the universe.

Plato

If you would like to make models of the Platonic solids, you will find cut-out diagrams at the back of this text.

As you can see in the following picture of the skeletons of radiolarians, the Platonic solids are not purely geometric in origin.

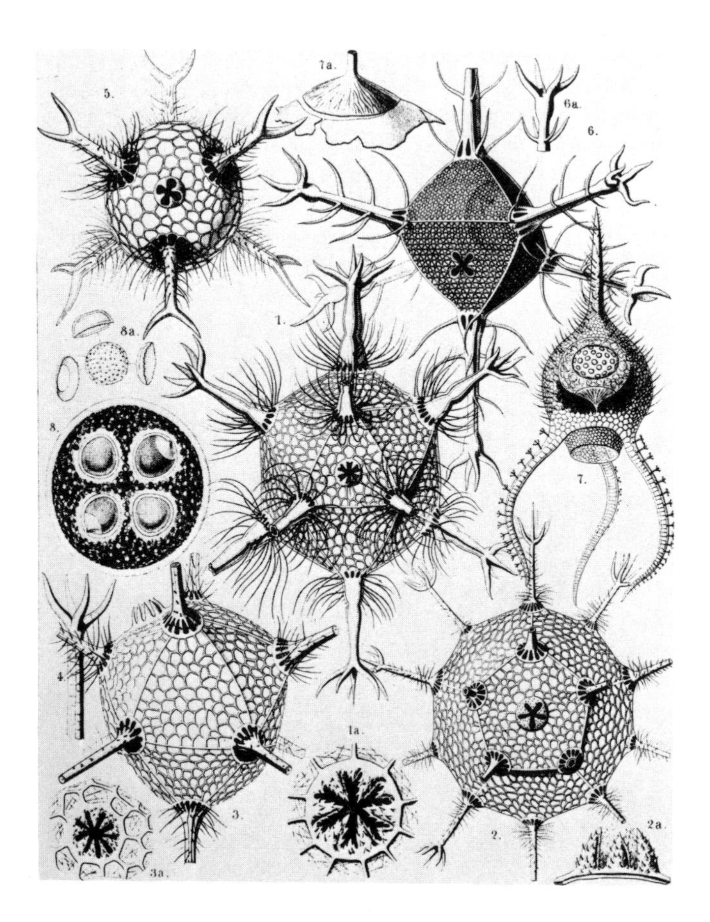

It is generally assumed that the first four Platonic solids at least were discovered from observations of crystals.

The Platonic solids clearly have reflectional *and* rotational symmetry. Let us consider the tetrahedron. We shall denote the vertices by V_1, V_2, V_3 and V_4, and the "centre" of the tetrahedron (equidistant from the four vertices) by O.

The tetrahedron has 6 distinct planes of symmetry (the planes of the triangles OV_iV_j $(i \neq j)$), 4 distinct three-fold axes of rotation (the lines OV_i), and 3 distinct two-fold axes of rotation (the bisectors of the angles V_iOV_j $(i \neq j)$). Even so, we have not mentioned *all* the mappings under which the tetrahedron is invariant! (There are 6 more, in addition to the identity mapping, each of which is the composite of a rotation about a line followed by a reflection in a plane perpendicular to that line.)

We leave you to count the planes and axes of symmetry for the other Platonic solids!

Now let's have a look at something a little different.

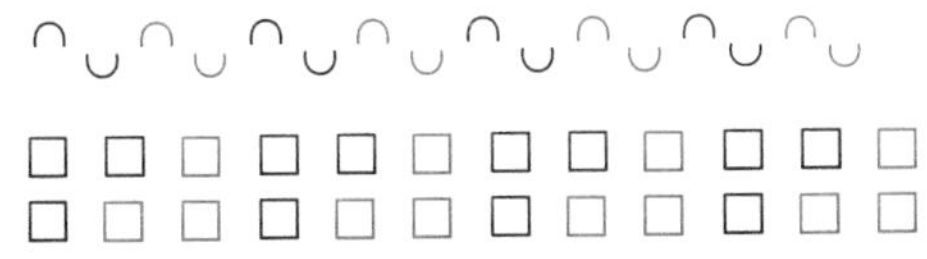

Several patterns similar to the above are used in the decoration of the Greek vase shown below.

The patterns illustrated differ from all our previous examples, because in each case the picture shows only a portion of a pattern which is meant to continue indefinitely across the plane. Under these circumstances there are new symmetry operations possible, namely *translations*. The patterns are unchanged if the plane is given a shift or translation of certain lengths in certain directions. Wallpaper and material patterns are good examples of (portions of) designs which exhibit translational symmetry; the cleverness of the design often lies in obscuring the translations. A good example of this is the decorative curtain in the National Film Theatre in London:

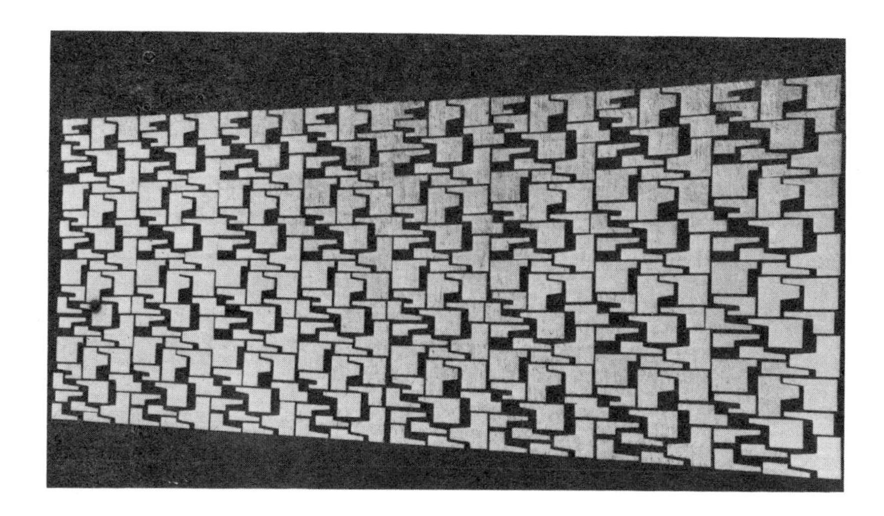

Notice that if a two-dimensional design admits a translation *and* a reflection or rotation, then the *image* under translation of a line or axis of symmetry must also be a line or axis of symmetry respectively. Furthermore, whereas a reflection composed with itself is the same as the identity mapping, under which each point in the domain is invariant, composing a translation with itself always produces a new, distinct symmetry operation of the pattern.

30.1.4 Algebraic Symmetry

30.1.4

Main Text
* *

When we express geometric symmetry in terms of mappings, we transfer from a geometric situation to an algebraic one. The fact that a circle centred at the origin is invariant under reflection in the y-axis has its algebraic counterpart in the fact that the equation

$$x^2 + y^2 - r^2 = 0$$

remains unchanged under the mapping

$$(x, y) \longmapsto (-x, y) \qquad ((x, y) \in R^2).$$

Equations and algebraic expressions can arise in non-geometric contexts, but we still refer to the idea of algebraic symmetry. For example, the mapping

$$\begin{cases} x_1 \longmapsto x_2 \\ x_2 \longmapsto x_1 \\ x_3 \longmapsto x_3 \\ x_4 \longmapsto x_4 \end{cases}$$

leaves the expression

$$x_1 x_2 + x_3 - x_4$$

unchanged. We say that the mapping is a *symmetry operation* on the particular expression; the mapping describes the symmetry of the expression.

There are many symmetry operations on the expression

$$x_1 + x_2 + x_3 + x_4.$$

(How many?) Some of these symmetry operations are also symmetry operations on the expression

$$(x_1 + x_2)(x_3 + x_4).$$

That is, the set of symmetry operations on the expression

$$(x_1 + x_2)(x_3 + x_4)$$

is a subset of the set of symmetry operations on the expression

$$x_1 + x_2 + x_3 + x_4.$$

We have seen that both geometric symmetry and algebraic symmetry can be expressed in terms of mappings. It is the structure that we obtain by performing these mappings successively that leads us to the idea of a group. In fact the concept of a group originated from the investigation of permutations and algebraic symmetry. The mathematician Galois (1811–1832) applied these ideas in tackling the problem of finding a formula for the roots of the equation

$$a_0 + a_1 x + a_2 x^2 + \cdots + a_n x^n = 0.$$

The details are not simple, but, roughly speaking, what Galois did was to associate a group with the coefficients $a_0, \ldots, a_n$, and then show that a formula for the roots could be found only if the group had certain properties. He transferred the problem from one in equations to one in group theory. At the same time he managed to apply the theory to prove the impossibility of a number of geometric constructions which had challenged mathematicians from ancient times. It was fortunate that he managed to record his results the night before he died (after a duel), at the age of twenty.

Évariste Galois

30.2 THE ALGEBRA OF SYMMETRY

30.2

30.2.0 Introduction

30.2.0

Introduction
* *

From the wealth of examples of symmetry, it would seem worth while to learn more about the properties of sets of symmetry operations. Such an investigation will lead us to discover four properties which are also exhibited in other contexts, and will motivate us to define the abstract concept of a *group*.

Mathematically, there is not very much of interest about a set such as a set of symmetry operations unless some binary operation is defined on it. Now when we sketched the graph with equation

$$|x| + |y| + |x + y| + |x - y| = 3$$

we used the important notion of *composition* of reflections, i.e. we followed one reflection by another; it is *composition* which enables us to combine symmetry operations.

In section 30.2.1 we shall examine the symmetry of the triquetrum and of the equilateral triangle. In section 30.2.2 we shall discuss the properties that are developed in section 30.2.1, and in section 30.2.3 we shall define the term *group*.

30.2.1 Two Examples

30.2.1

Discussion
* *

The types of symmetry operation we have mentioned so far have all been described in terms of mappings of the space containing some given figure to itself. If we are going to allow all possible symmetry operations, then we must be more precise about what sort of mappings we are going to admit. Now our notion of symmetry is based on our physical experience, and the kinds of mapping that conform to our physical experience have the property that they do not change either lengths or angles; such mappings are called "rigid" transformations, since they preserve the distance between any two points in their domains. Formally then, a symmetry operation applied to a figure in space is a "rigid" mapping of the space to itself which also maps the figure to itself, that is, under which the figure is invariant.

Definition 1
* * *

When we introduced rotations, we discovered the need to be able to tell when two symmetry operations have the same overall effect. When they do, we want to treat them as the same symmetry operation, so another concept which we want to specify precisely is what we mean by the *equality* of symmetry operations.

In the case of rotations, we equated rotations through 0 and 2π because in both cases the general point (x, y) maps to itself. Similarly, rotations through π and 3π can be equated, since in both cases $(x, y) \longmapsto (-x, -y)$. It would seem at first that a suitable definition of equality might be that, if S and T are symmetry operations, then $S = T$ if and only if S and T have the same effect on every point of the space. But we are not interested in the surrounding space: we are interested in the symmetry of a particular figure. For instance, consider a line-segment lying in the plane of the page as shown. We can reflect the plane in an axis through the line-segment, and although no point of the line-segment moves, the points of the plane certainly do.

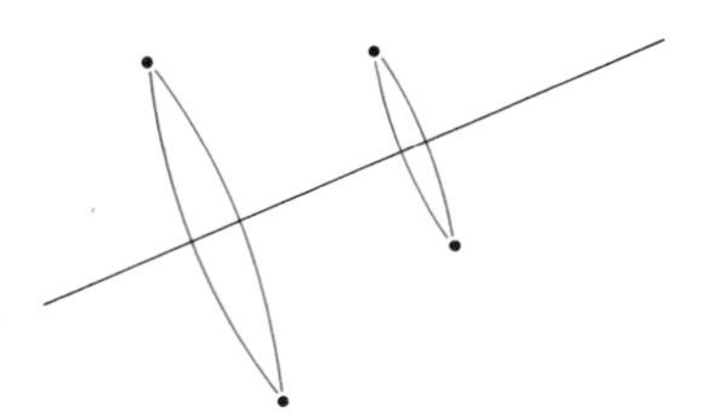

In this case, we are interested in the symmetry of the line-segment, so we want to equate this particular reflection with the symmetry operation which maps *every* point to itself, since the two have the same effect on the line-segment. Therefore, we define two symmetry operations S and T on a figure (or algebraic expression) to be equal if and only if the image of each point of the figure (or the image of the expression) is the same under S as under T. If S and T are equal, we write $S = T$.

Definition 2
* * *

There is a simple picture which may help in visualizing geometric symmetry operations. Think of the given figure as lying in a tight-fitting compartment of a box. Suppose that we lift the figure out of the box, perform a symmetry operation on it and then replace it. In order to find out whether two symmetry operations are the same, we attach labels to salient points of the figure before we apply a symmetry operation and then observe their new positions.

For example, let us analyse the symmetry of the triquetrum. First we show a picture of the labelled triquetrum in its box.

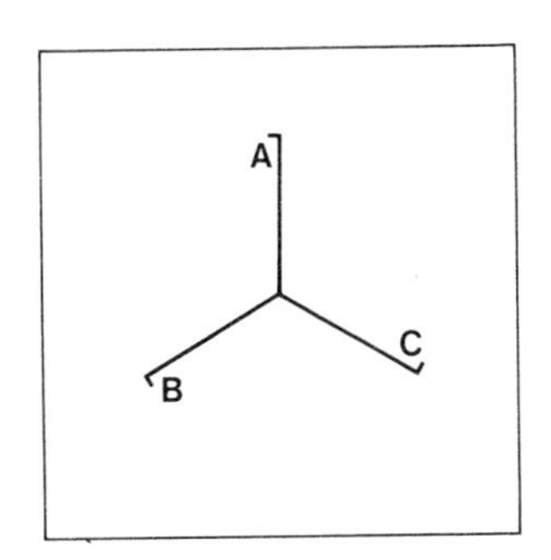

Now when we apply a symmetry operation, we shall just be picking up the triquetrum and replacing it in the box. Two symmetry operations are equal if the labels end up in the same places when we apply either of the symmetry operations to the same starting position. Thus, distinct symmetry operations correspond to distinct ways of putting the labelled triquetrum back in its box. The next pictures show the three possible ways of putting the triquetrum back in its box, starting from the central position.

It is common practice to denote a symmetry operation which does not change any point of the figure by the letter e. In the case of the triquetrum we have called the other symmetry operations T and S for no particular reason. Notice that e, T, S correspond to clockwise rotations through angles 0, $\frac{2\pi}{3}$ and $\frac{4\pi}{3}$ respectively about an axis perpendicular to the plane of the triquetrum through its centre.

Now the operation on the set of mappings $\{e, T, S\}$ is to be composition of mappings. Remember that we write $S \circ T$ to mean the result of first applying T and then applying S. Starting with a triquetrum labelled as on the left, we can readily calculate the result of $S \circ T$.

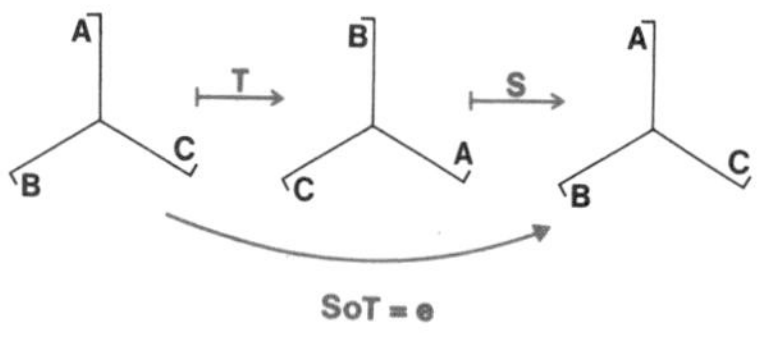

A B C T B C A S A B C
SoT = e

Notice that the end result is the same as just applying e to the initial figure and that this is so no matter what initial position we choose; that is, $S \circ T = e$. We can consider in turn each of

$$e \circ S,\ e \circ T,\ S \circ e,\ T \circ e,\ S \circ T,\ T \circ S,\ e \circ e,\ S \circ S,\ T \circ T.$$

It is convenient to display the results in a table as shown. The convention is that $S \circ T$ is found at the intersection of the row labelled S and the column labelled T. (Notice that, in this case, we have a symmetric table: composition of rotations is commutative.)

$\circ$	e	T	S
e	e	T	S
T	T	S	e
S	S	e	T

In the table we see that every entry is one of the original symmetry operations. Is it obvious that the composition of two symmetry operations is again a symmetry operation? By definition, each symmetry operation maps the space to itself and leaves the figure invariant. Thus if we perform two symmetry operations in succession, the overall effect is still to map the space to itself and leave the figure invariant, and hence the composite is itself a symmetry operation.

Notice also that each symmetry operation appears once and once only in each row and each column of the table.

As a second example, we look at the symmetry of an equilateral triangle. Again, think of the triangle sitting in its box. Label the salient points (say the vertices), and then list the various ways of putting it back in its box. There are 3 distinct rotations about the centre of the triangle, and 3 distinct reflections in the medians (shown by dotted lines in the following figure).

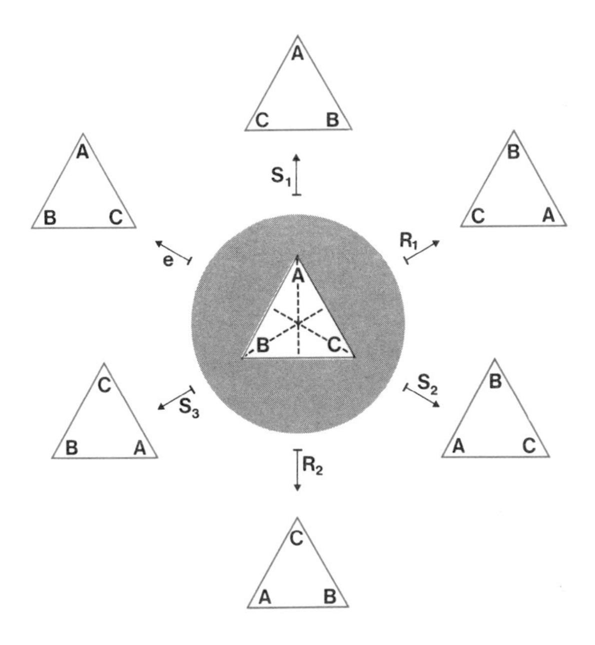

We have again labelled the "no change" (identity) symmetry operation by e. The clockwise rotations through angles $\frac{2\pi}{3}$ and $\frac{4\pi}{3}$ are labelled R_1 and R_2 respectively, and the reflections in the medians are labelled S_1, S_2, S_3. The rotations and reflections are, however, inter-related, as we see when we write down the results of the 36 possible compositions. This time a table is absolutely essential!

$\circ$	e	R_1	R_2	S_1	S_2	S_3
e	e	R_1	R_2	S_1	S_2	S_3
R_1	R_1	R_2	e	S_3	S_1	S_2
R_2	R_2	e	R_1	S_2	S_3	S_1
S_1	S_1	S_2	S_3	e	R_1	R_2
S_2	S_2	S_3	S_1	R_2	e	R_1
S_3	S_3	S_1	S_2	R_1	R_2	e

Notice that the red block of the table is the same as the table for the triquetrum; we have used R_1 instead of T and R_2 instead of S. The rest of the table is obtained in a similar manner. For example, if we perform the same reflection twice, the triangle will be unchanged. That is,

$$S_1 \circ S_1 = S_2 \circ S_2 = S_3 \circ S_3 = e.$$

Here the composite of two reflections is a rotation and the composite of a rotation and a reflection is a reflection. Note that, in this case, composition is *not* commutative: for example,

$$R_1 \circ S_2 = S_1$$

$$S_2 \circ R_1 = S_3$$

so

$$R_1 \circ S_2 \neq S_2 \circ R_1.$$

Exercise 1

Exercise 1
(4 minutes)

Write down the symmetry table for:

(i) a rectangle;
(ii) a square. ■

Solution 1

Solution 1

(i) *Rectangle*

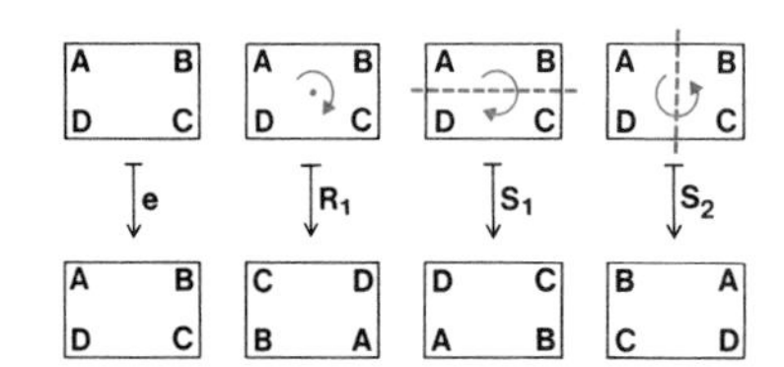

Labelling the "no change" operation by e, the clockwise rotation R_1 and the two reflections S_1 and S_2, as shown, we get the following table.

$\circ$	e	R_1	S_1	S_2
e	e	R_1	S_1	S_2
R_1	R_1	e	S_2	S_1
S_1	S_1	S_2	e	R_1
S_2	S_2	S_1	R_1	e

(ii) *Square*

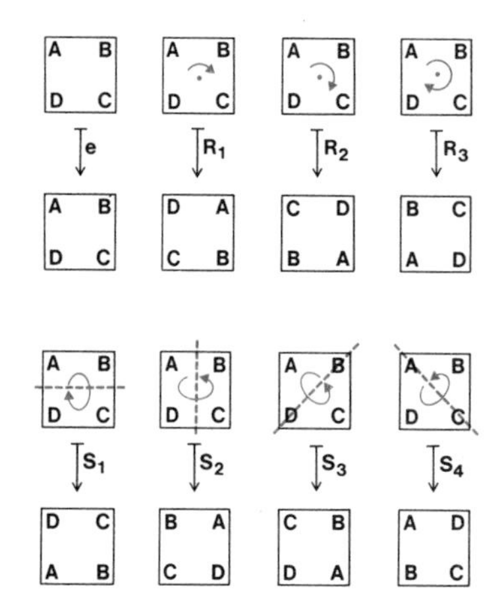

Labelling the "no change" operation by e, the clockwise rotations through $\frac{\pi}{2}$, π, $\frac{3\pi}{2}$ by R_1, R_2 and R_3 respectively and the reflections by S_1, S_2, S_3 and S_4, as shown, we get the table given below.

$\circ$	e	R_1	R_2	R_3	S_1	S_2	S_3	S_4
e	e	R_1	R_2	R_3	S_1	S_2	S_3	S_4
R_1	R_1	R_2	R_3	e	S_4	S_3	S_1	S_2
R_2	R_2	R_3	e	R_1	S_2	S_1	S_4	S_3
R_3	R_3	e	R_1	R_2	S_3	S_4	S_2	S_1
S_1	S_1	S_3	S_2	S_4	e	R_2	R_1	R_3
S_2	S_2	S_4	S_1	S_3	R_2	e	R_3	R_1
S_3	S_3	S_2	S_4	S_1	R_3	R_1	e	R_2
S_4	S_4	S_1	S_3	S_2	R_1	R_3	R_2	e

Again, the red block of the table embodies the results of combining pairs of rotations. For example, a rotation through $\frac{\pi}{2}$ and then a rotation through π is equivalent to a rotation through $\frac{3\pi}{2}$; that is,

$$R_2 \circ R_1 = R_3.$$

Again the composite of two reflections is a rotation, and the composite of a rotation and a reflection is a reflection. ■

30.2.2 Properties of Symmetry Tables

30.2.2

Main Text
* * *

In this section we shall extract certain properties that are exhibited by the four symmetry tables in the previous section (triquetrum, equilateral triangle, rectangle, square). The properties are shared by all symmetry tables, and will lead us to the more general concept of a *group*.

(i) The composite of two symmetry operations is again a symmetry operation. In other words the set of symmetry operations on a figure is *closed* under the operation of composition.
(ii) The symmetry operation e has the property that

$$S \circ e = e \circ S = S,$$

for every symmetry operation S, and *always* belongs to the set of all symmetry operations on some figure.
(iii) Each row of the table contains each symmetry operation precisely once, and so does each column.

These observations, as we shall see, have many consequences.

First, it follows that the symmetry operation e occurs just once in each row and column. That means that for any symmetry operation a, we can find some symmetry operation b such that $a \circ b = e$, and a symmetry operation c such that $c \circ a = e$. In fact, we can observe from our tables that, at least in those four cases, $b = c$. We shall be able to prove this in general, once we have established a set of axioms on which we can base proofs.

A second consequence is that we can think of any one symmetry operation, a say, as defining a mapping f from the set of symmetry operations $\mathscr{S}$ to itself. For example, we could use the symmetry operation a to define the mapping

$$f: S \longmapsto a \circ S \qquad (S \in \mathscr{S}).$$

For example, in the table for the equilateral triangle, the symmetry operation $a = R_1$ defines the mapping under which

$$e \longmapsto R_1 \circ e = R_1$$
$$R_1 \longmapsto R_1 \circ R_1 = R_2$$
$$R_2 \longmapsto R_1 \circ R_2 = e$$
$$S_1 \longmapsto R_1 \circ S_1 = S_3$$
$$S_2 \longmapsto R_1 \circ S_2 = S_1$$
$$S_3 \longmapsto R_1 \circ S_3 = S_2.$$

The fact that each symmetry operation occurs exactly once in row a of the table tells us that the mapping associated with a is a one-one function of the set of symmetry operations to itself (i.e. the image set is the same as the domain). Since a one-one function can be regarded as just a rearrangement of the set of symmetry operations, it is usually called a permutation. In the example of the equilateral triangle, the list

Definition 1
* * *

$$e, R_1, R_2, S_1, S_2, S_3$$

was rearranged by the permutation to give

$$R_1, R_2, e, S_3, S_1, S_2.$$

One property that is not obvious from the table is that *composition is associative**. We did, however, show in *Unit 23, Linear Algebra II*, section 23.2.3, that composition of composable functions is an associative binary operation.

Two further observations, while not of fundamental importance to the process of abstracting the notion of a group, nevertheless contain ideas essential to the study of symmetry and groups.

(i) The subset $\{e\}$ of the set of symmetry operations of a figure is itself closed under composition, since $e \circ e = e$. The subset $\{e, R_1, R_2\}$ of the set of symmetry operations of the equilateral triangle is also closed under composition.
(ii) For the triquetrum, if we were just allowed to use the symmetry operation T, we could, by means of composition of T with itself, produce the full set of symmetry operations, since $T \circ T = S$ and $T \circ (T \circ T) = e$. The symmetry operation T is said to *generate* the set of symmetry operations of the triquetrum.

If we try to do the same for the equilateral triangle, we quickly find that there is no one symmetry operation which will generate all the rest, but we can pick two, which together will generate the full set. That is, if we choose S_1 and R_1, say, then

$$R_1 \circ R_1 = R_2$$
$$R_1 \circ (R_1 \circ R_1) = e$$
$$R_1 \circ S_1 = S_3$$
$$S_1 \circ R_1 = S_2$$

and so together S_1 and R_1 generate the set of symmetry operations.

More generally, we say that a subset $\mathcal{S}_1$ of the set of symmetry operations $\mathcal{S}$ of a figure generates the full set $\mathcal{S}$, if for every $S \in \mathcal{S}$ we can express S as a composition of elements of $\mathcal{S}_1$. One way of studying sets of symmetry operations, and, more generally, groups, is by studying subsets which generate them. For example, the whole set of symmetry operations of the equilateral triangle is characterized by the fact that there are two symmetry operations, S_1 and R_1, with the properties that

Definition 2
* *

$$R_1 \circ R_1 \circ R_1 = e$$

and

$$S_1 \circ S_1 = e,$$

and that S_1 and R_1 generate the whole set.

Exercise 1

Exercise 1
(3 minutes)

Find all the proper subsets of the sets of symmetry operations of the triquetrum, equilateral triangle, rectangle and square, which are closed under composition. ■

* There is a visual way of testing for associativity in a group table. If you are interested, see M. Bruckheimer and J. V. Scott, "Testing for Associativity", in *Mathematics Teaching*, 45 (1968), 44.

Exercise 2

Exercise 2
(4 minutes)

What is the minimum number of symmetry operations needed to generate the set of symmetry operations of

(i) the rectangle;
(ii) the square? ■

Discussion
* *

It took mathematicians a long time to digest the observations we have just made, and to organize them in a coherent way so that a real study of symmetry could be made. The first major step is to isolate four fundamental properties which we can treat as axioms. We can then proceed to deduce information about symmetry operations, or indeed about any set which satisfies the axioms.

30.2.3 The Concept of a Group

30.2.3

Main Text
* * *

Of all the remarks which we made in the last section concerning the properties of the symmetry tables, the following four might be considered as basic: they can be taken as the starting point for investigating all the other properties.

(i) If a and b are symmetry operations, then $a \circ b$ is a symmetry operation (closure).
(ii) If a, b and c are symmetry operations, then

$$(a \circ b) \circ c = a \circ (b \circ c)$$

(associativity).
(iii) There is a symmetry operation e such that

$$a \circ e = a = e \circ a$$

for *every* symmetry operation a
(e is called the *identity* symmetry operation).
(iv) Given *any* symmetry operation a, we can always find a symmetry operation b for which

$$a \circ b = e$$

(b is called an *inverse* of a).

So the four ideas are closure, associativity, identity element and inverse element. None of them is new; we have seen many examples of each idea earlier in the course. We now put the four ideas together and examine the new structure.

Definition 1
* * *

We now define a group $(G, \circ)$ to be a set G with a binary operation $\circ$ defined on it, with the following properties:

(i) $\circ$ is closed;
(ii) $\circ$ is associative;
(iii) there is an *identity element* $e \in G$ such that for all $a \in G$

$$a \circ e = a = e \circ a;$$

(iv) for any element $a \in G$, there is an *inverse element* $b \in G$ such that

$$a \circ b = e.$$

We shall show later that there is only *one* identity element in any group, and that each element of the group has a *unique* inverse.

Definition 2
* * *

If G has a finite number of elements, n say, we say we have a finite group of order n: otherwise we have a group of infinite order.

So far we have considered groups of symmetry operations with the binary operation of composition.

(*continued on page 29*)

Solution 30.2.2.1

Solution 30.2.2.1

Triquetrum Only the trivial subset $\{e\}$.

Equilateral Triangle $\{e, R_1, R_2\}$, $\{e, S_1\}$, $\{e, S_2\}$, $\{e, S_3\}$, $\{e\}$.

Rectangle Using the notation of Solution 30.2.1.1, the following subsets are closed:

$$\{e, R_1\}, \{e, S_1\}, \{e, S_2\}, \{e\}.$$

Square Again using the notation of Solution 30.2.1.1, we get:

$$\{e, R_1, R_2, R_3\}, \{e, R_2, S_1, S_2\}, \{e, R_2, S_3, S_4\}, \{e, R_2\},$$
$$\{e, S_1\}, \{e, S_2\}, \{e, S_3\}, \{e, S_4\}, \{e\}.$$

■

Solution 30.2.2.2

Solution 30.2.2.2

(i) To generate the set of symmetry operations of the rectangle, we need at least two symmetry operations. For example, S_1 and R_1 would serve the purpose, for we can obtain the other two elements as follows

$$S_1 \circ S_1 = e$$
$$S_1 \circ R_1 = S_2.$$

(ii) The set of symmetry operations of the square can also be generated using two symmetry operations, R_1 and S_1 for example.

We have

$$R_1 \circ R_1 = R_2$$
$$R_1 \circ R_1 \circ R_1 = R_3$$
$$R_1 \circ S_1 = S_4$$
$$S_1 \circ R_1 = S_3$$
$$R_1 \circ R_1 \circ S_1 = S_2$$
$$S_1 \circ S_1 = e.$$

■

We shall now consider other sets, and other operations. As usual the easiest things to think about first are numbers. We have all sorts of operations on numbers; let us start with addition. Certainly the set of real numbers is closed under addition, and addition is an associative operation. What about an identity? Is there a real number e such that

$$a + e = a = e + a$$

for *every* real number a? Obviously we have an identity element: $e = 0$. And inverses? Given any real number a, can we find a real number x such that

$$a + x = 0?$$

The inverse of a is $-a$ for all $a \in R$.

So the set of real numbers under the operation of addition has the same four properties as our sets of symmetry operations under the operation of composition. That is, $(R, +)$ *is a group*. And now the whole topic becomes more significant. Probably the best measure of the importance of a piece of mathematics is its generality: the scope of its application. The great mathematician G. H. Hardy wrote* "... a mathematical idea is significant if it can be connected, in a natural and illuminating way, with a large complex of other mathematical ideas". On this count, the ideas we have here are among the most significant in mathematics. The importance is realized when we isolate the four properties of symmetry operations, strip them of their external association and consider the abstract mathematical structure thus revealed.

Example 1

Example 1

Consider the real numbers again, this time with the operation of multiplication. The identity is 1 and the inverse of a is $1/a$, $a \neq 0$. Because we must have an inverse for every element, we must in this case exclude the number 0 from the set. We know that multiplication on the set of non-zero real numbers is closed and associative. That is, $(R^+ \cup R^-, \times)$ is a group. ■

Exercise 1
(3 minutes)

Exercise 1

Investigate whether the following are groups:

(i) the set of rational numbers under
   (a) addition, (b) multiplication;
(ii) the set of natural numbers, $\{1, 2, 3, \ldots\}$, under
   (a) addition, (b) multiplication. ■

Exercise 2
(3 minutes)

Exercise 2

In each of the following cases confirm that the structure is a group.

(i) The set of complex numbers under addition.
(ii) The set of complex numbers, without 0, under multiplication.
(iii) The set of all $m \times n$ matrices (for some given m and n) under matrix addition.
(iv) The set of all translations of the points in a plane, under composition of translations. ■

Exercise 3
(4 minutes)

Exercise 3

(i) Suggest a suitable set of matrices which will give a group under matrix multiplication.
(ii) For the set of all geometric vectors in a plane, do we get a group for (a) vector addition, (b) inner product, (c) multiplication by a scalar?
(iii) Show that, for given n, the set of all solutions of the equation $z^n = 1$ forms a group for multiplication of complex numbers. ■

* G. H. Hardy, *A Mathematician's Apology* (Cambridge University Press, 1967).

Solution 1

Solution 1

(i) The set of all rationals forms a group for addition, with 0 as the identity and $-p/q$ as the inverse of p/q. The set of all non-zero rationals forms a group for multiplication. For multiplication, 1 is the identity, q/p is the inverse of p/q, $p \neq 0$, $q \neq 0$. The number 0 has no inverse, and so we do not get a group unless zero is excluded. The product of two non-zero rationals is non-zero, so multiplication *is* a closed operation.

(ii) This set does not form a group for either operation. For addition there is no identity (and the question of inverses does not therefore arise). For multiplication, 1 is an identity, but the inverses do not exist. ■

Solution 2

Solution 2

(i) Addition of complex numbers is closed and associative.
The identity element is 0.
The inverse of z is $-z$.
(See *Unit 27, Complex Numbers I*.)

(ii) Multiplication of non-zero complex numbers is closed and associative.
The identity element is 1.
The inverse of z ($z \neq 0$) is $1/z$.
(See *Unit 27*.)

(iii) Matrix addition of $m \times n$ matrices is closed and associative.
The identity element is the $m \times n$ null matrix.
The inverse of A is $-A$.
(See *Unit 23*.)

(iv) Composition of translations is closed and associative.
The identity is the zero translation, corresponding to $\underset{\sim}{0}$.
The inverse of the translation corresponding to $\underset{\sim}{a}$ is the translation corresponding to $-\underset{\sim}{a}$.
(See *Unit 22*.)

Solution 3

Solution 3

(i) The set of all matrices of the form $\begin{bmatrix} \alpha & 0 \\ 0 & \alpha \end{bmatrix}$, $\alpha \neq 0$, is one such set.

$$\begin{bmatrix} \alpha & 0 \\ 0 & \alpha \end{bmatrix}\begin{bmatrix} \beta & 0 \\ 0 & \beta \end{bmatrix} = \begin{bmatrix} \alpha\beta & 0 \\ 0 & \alpha\beta \end{bmatrix}$$

and $\alpha \neq 0$ and $\beta \neq 0 \Rightarrow \alpha\beta \neq 0$, so multiplication is closed, and we know that matrix multiplication is associative. (See *Unit 23*.)

The identity is $\begin{bmatrix} 1 & 0 \\ 0 & 1 \end{bmatrix}$.

The inverse of $\begin{bmatrix} \alpha & 0 \\ 0 & \alpha \end{bmatrix}$ is $\begin{bmatrix} 1/\alpha & 0 \\ 0 & 1/\alpha \end{bmatrix}$.

(ii) (a) The set of geometric vectors is a group under vector addition. The axioms of an abstract vector space include the group axioms with respect to vector addition.

(b) For the inner product we do not get closure: the inner product of two vectors is a scalar, not a vector.

(c) This is not even a binary operation on the set of vectors.

(iii) If z_1 and z_2 are two solutions of the equation, then

$$z_1^n = 1 \quad \text{and} \quad z_2^n = 1.$$

But $(z_1 z_2)^n = z_1^n z_2^n = 1$, so multiplication is closed.

It is associative, because the multiplication of complex numbers is associative. The number 1 is a solution and acts as the identity for the set. Also if

$$z_1^n = 1, \quad \text{then } z_1 \neq 0 \quad \text{and} \quad \left(\frac{1}{z_1}\right)^n = \frac{1}{z_1^n} = 1.$$

So for each z_1 there is an inverse $\dfrac{1}{z_1}$ in the set. ■

Discussion
* *

The fact that, as we have just seen, there are examples of groups in the ordinary arithmetic of numbers, suggests that groups may well underlie much of the mathematics that we have already met. This is reinforced by the examples in Exercises 2 and 3. Of course, the real numbers have a much richer structure than a group — for a start they have more than one operation defined on them, but that is not to suggest that the idea of a group is feeble. It is often the case that one wants to simplify a structure in some sense: to discuss just one facet of that structure. This is the sort of procedure we adopted in our development of linear algebra. When we discussed geometric vectors we defined two basic operations: vector addition and the inner product. The concepts of length, angle and distance are all bound up with the inner product, but we shelved these when we developed geometric vectors to abstract vector spaces. We took just vector addition and multiplication by a scalar and saw where these led us. And, as we have just seen in Exercise 3, vectors under vector addition form a group. So what we did was essentially to pick out a group from the structure of geometric vectors and develop the idea of dimension and so on using just this group structure, combined with the real numbers. It was a special kind of group in the sense that vector addition is commutative, and we do not require commutativity in the general definition of a group.

Thus a vector space is essentially a (commutative) group combined in some way with the real numbers (or perhaps another structure with similar properties to the real numbers).

We are at an interesting stage in the development of group theory. From this point onward one can regard the study of groups in several lights. We have defined the abstract structure of a group; it is a set with an operation which satisfies four particular properties. We could take this as the starting point of the development of a piece of abstract mathematics. Alternatively, one might be prompted to investigate groups because one's field of interest leads that way. For example, crystallographers have an interest in group theory, and they study the subject to help them with their own work. They may not find group theory gripping as a piece of pure mathematics, but they find it a most useful tool. And there is another class of person who may be interested in the universality of groups; because they touch on so much mathematics, a study of groups can give a rewarding insight into the similarities, differences, and links between various branches of mathematics. The idea of a group is also of interest to psychologists. For example, there is a school of thought which holds that there is a clearly marked stage in a child's development when the idea of an inverse manifests itself.

Whatever attitude one adopts, it is surely quite remarkable that so much should follow from these four simple axioms.

30.2.4 Further Examples

30.2.4

Examples
* *

Before we develop the axioms and make some remarks about groups in general, we shall give some more examples of groups. We shall also introduce the concept of a *subgroup*.

Exercise 1

Exercise 1
(5 minutes)

Which of the following tables represents a group? State one axiom which fails to hold if the table does not represent a group.

(i)

$\circ$	A	B	C	D
A	D	C	B	A
B	C	D	A	B
C	B	A	D	C
D	A	B	C	D

(ii)

$\circ$	A	B	C	D
A	A	B	C	D
B	D	A	B	C
C	C	D	A	B
D	B	C	D	A

(iii)

$\circ$	A	B	C	D
A	A	B	C	D
B	B	A	A	D
C	D	A	C	D
D	B	C	D	C

(iv)

$\circ$	A	B	C	D	E
A	E	C	D	B	A
B	C	D	A	E	B
C	D	E	B	A	C
D	B	A	E	C	D
E	A	B	C	D	E

(*Note* If you think that, because the letter e does not occur, these tables cannot represent groups, remember that we generally agreed to *call* the identity element e, but it doesn't matter what it is called. See if some other element is playing the role of e in each of these tables.) ■

Discussion
* *

In part (i) of Exercise 1 we have a group table which is essentially the same as the group table of the rectangle: the only difference is that we have used different letters. To all intents and purposes we have the same group. How can we identify groups which are essentially different?

One of the first problems in group theory is the classification and enumeration of different groups.

A first step is to state the number of elements in the group. We call this number the *order* of the group. Thus the groups of symmetry operations of the triquetrum, equilateral triangle, rectangle and square have orders 3, 6, 4 and 8 respectively, and the set of real numbers under addition is a group of infinite order. We can have a group of order any natural number. For instance, the group of symmetry operations of an n-armed pin-wheel has order n; there are precisely n possible rotations : rotations through

$$0, \frac{2\pi}{n}, \frac{4\pi}{n}, \frac{6\pi}{n}, \ldots, \frac{2(n-1)\pi}{n}.$$

n = 1

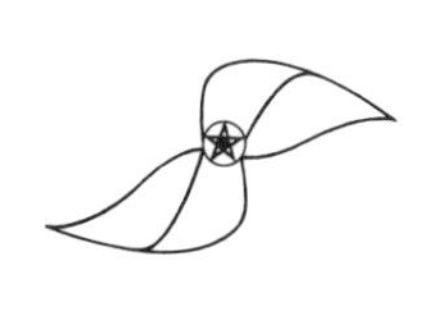

n = 2

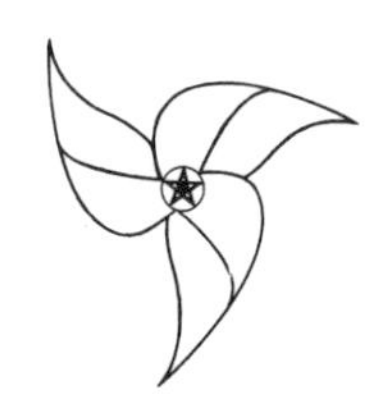

n = 3

n-armed pin-wheels

The next question is: for a given integer n, how many different groups are there of order n? The answer is not known in the general case. Of course, we must make precise what we mean by "different" groups, and we shall do this in *Unit 33, Group II* when we talk about *morphisms* from one group to another. We can however, observe that the groups of symmetry operations of a rectangle and of a four-armed pin-wheel, which are shown below, are, in any reasonable sense of the word, "different". One reason is that the group of the rectangle requires two symmetry operations to generate it, whereas the group of the pin-wheel is generated by just one symmetry operation, and so the two groups have different structure. Notice also that for the rectangle, $g \circ g = e$ no matter which element g of the group we choose, but this is not the case for the pin-wheel group.

$\circ$	e	R	S	T
e	e	R	S	T
R	R	e	T	S
S	S	T	e	R
T	T	S	R	e

Rectangle group

$\circ$	e	A	B	C
e	e	A	B	C
A	A	B	C	e
B	B	C	e	A
C	C	e	A	B

4-armed pin-wheel group

The "pin-wheel" group is an example of a frequently occurring type of group called a cyclic group; it is generated by a single element, and is characterized by rotational symmetry. In general, the group of rotations through angles $0, \frac{2\pi}{n}, \frac{4\pi}{n}, \frac{6\pi}{n}, \ldots, \frac{2(n-1)\pi}{n}$ is called the cyclic group of order n and denoted by C_n.

Definition 1
* * *

Notation 1
* * *

The rotational symmetry operations of the flower illustrated below form a cyclic group of order 5, C_5.

In Exercise 30.2.3.3 we had an example of a cyclic group of order n: the set of complex roots of the equation

$$z^n = 1,$$

with the operation of multiplication of complex numbers.

(*continued on page 34*)

Solution 1

Solution 1

(i) This table represents a group. If we relabel according to the mapping

$$A \longmapsto S_2$$
$$B \longmapsto S_1$$
$$C \longmapsto R_1$$
$$D \longmapsto E$$

the table becomes exactly the same as the one we obtained in Exercise 30.2.1.1 for the symmetry operations of the rectangle.

(ii) This table does not represent a group. There is no identity; although A looks a candidate (see top line of table), it is only an identity on the left. When it is placed on the right, we have, for instance,

$$B \circ A \neq B.$$

(iii) This table does not represent a group. There is no identity element.

(iv) This table does not represent a group, E is an identity element, the operation is closed and inverses all exist, but the operation is not associative; for example,

$$A \circ (B \circ C) = A \circ A = E$$

but

$$(A \circ B) \circ C = C \circ C = B.$$

■

(*continued from page 33*)

A cyclic group also arises from the symmetry operations of the molecule of hydrogen peroxide. This molecule is made up of two hydrogen atoms and two oxygen atoms arranged symmetrically as in the diagram.

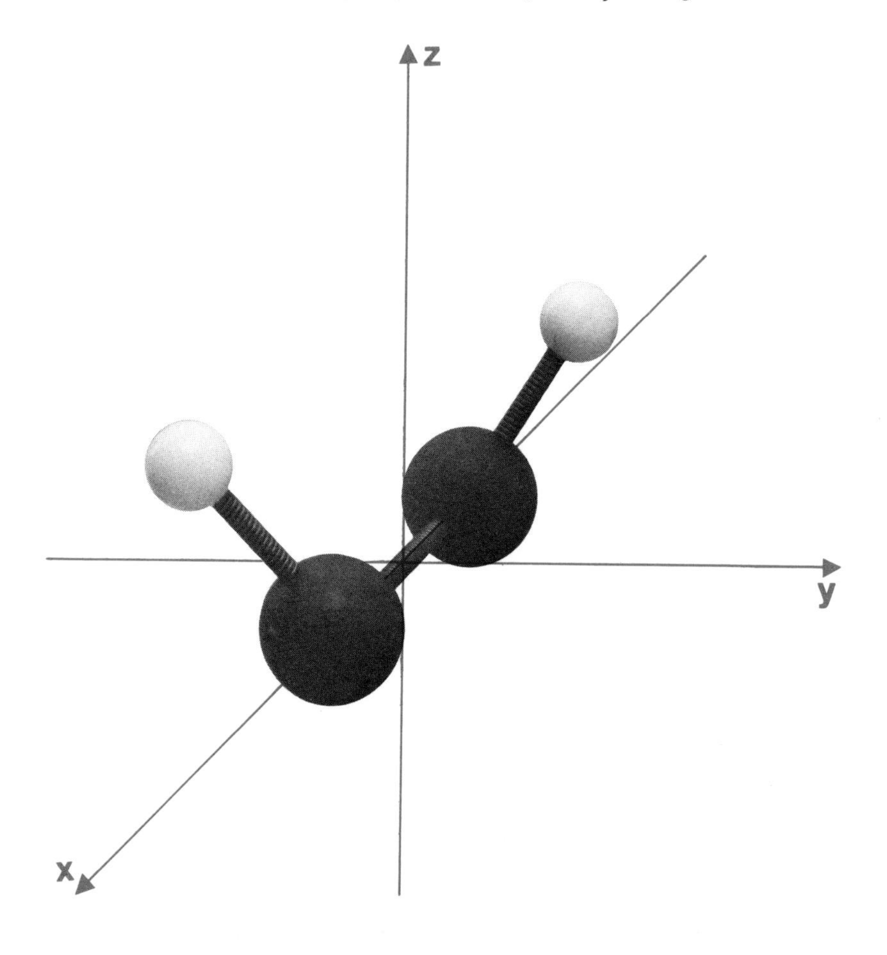

The molecule is mapped to itself under rotations of 0 and π about the z-axis, and so the symmetry operations of the molecule form a cyclic group of order 2. We have the following group table:

$\circ$	e	R
e	e	R
R	R	e

where R denotes a rotation through π about the z-axis.

A different type of symmetry is exhibited by the water molecule. This is a planar molecule; it contains two hydrogen atoms and one oxygen atom arranged like this.

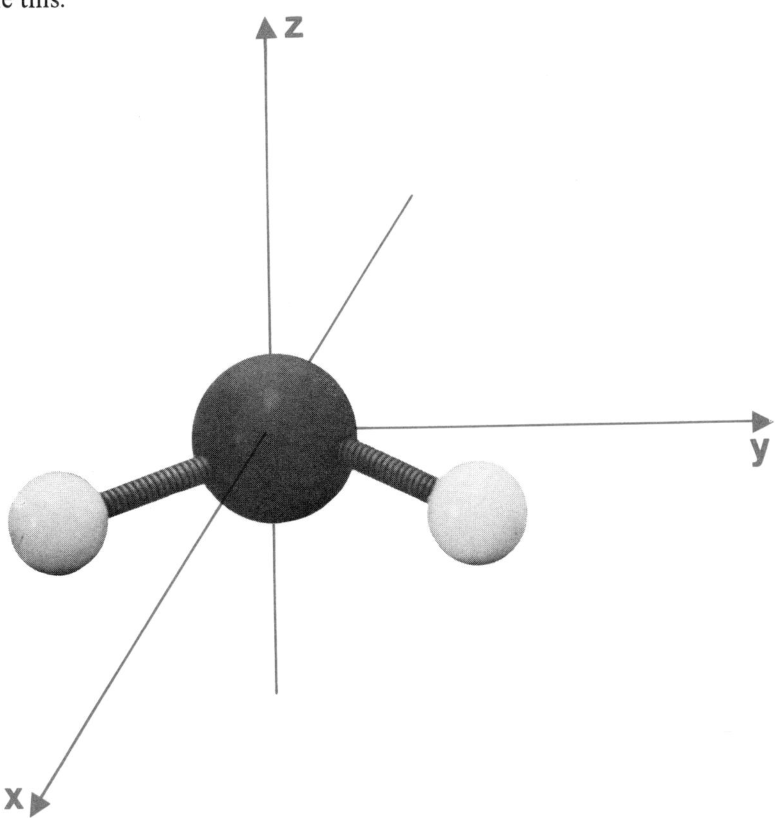

As with the hydrogen peroxide molecule, this molecule has rotational symmetry operations through 0 and π (about the x-axis). We can also reflect the molecule in the xz-plane, and we can reflect it in the xy-plane containing the three atoms. This molecule has the following group table:

$\circ$	e	R_1	S_1	S_2
e	e	R_1	S_1	S_2
R_1	R_1	e	S_2	S_1
S_1	S_1	S_2	e	R_1
S_2	S_2	S_1	R_1	e

where we use S_1 and S_2 to stand for the reflections, and R_1 to stand for the rotation through π. We have met this group before: it is the symmetry group of the rectangle. (Note that when we considered the symmetry of the rectangle, etc., we did not count the plane of the figure as a plane of symmetry — it was of no interest. However, to a chemist, the fact that a molecule is planar *is* of interest; it tells him, for example, that its vibrations are symmetric about its plane; so a chemist calls this plane a plane of symmetry and includes the corresponding reflection in the group.) You will see the symmetry group of the water molecule playing a large part in the television programme associated with this unit. It is called the Klein 4-group (after the German mathematician Klein (1849–1925)).

Felix Klein

A rather more complicated example is the molecule of methyl chloride. This contains three hydrogen atoms, one carbon atom and one atom of chlorine arranged as in the diagram below.

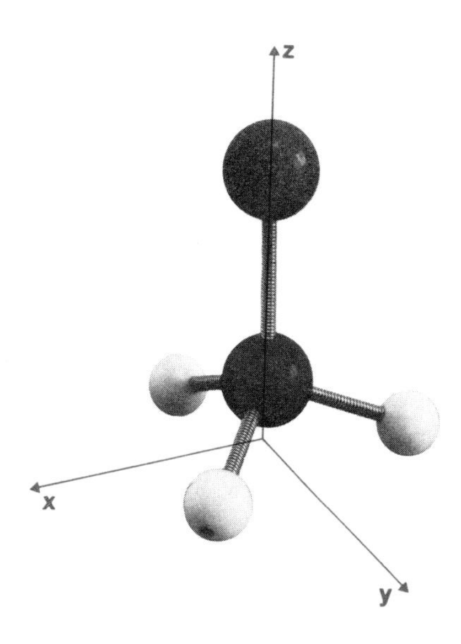

The hydrogen atoms are positioned at the vertices of an equilateral triangle. The carbon and chlorine atoms are on a line which passes through the centre of this triangle. So the molecule has a three-fold axis of symmetry (the z-axis) and it also has 3 vertical planes of symmetry (each containing the z-axis and a C—H bond).

Its group table is the same as the symmetry group for the equilateral triangle which we discussed in section 30.2.1.

$\circ$	e	R_1	R_2	S_1	S_2	S_3
e	e	R_1	R_2	S_1	S_2	S_3
R_1	R_1	R_2	e	S_3	S_1	S_2
R_2	R_2	e	R_1	S_2	S_3	S_1
S_1	S_1	S_2	S_3	e	R_1	R_2
S_2	S_2	S_3	S_1	R_2	e	R_1
S_3	S_3	S_1	S_2	R_1	R_2	e

R_1 and R_2 denote the rotations through $\frac{2\pi}{3}$ and $\frac{4\pi}{3}$ respectively about the z-axis; S_1, S_2 and S_3 denote the reflections in the three planes of symmetry.

So we see that we can discuss the symmetry of molecules in terms of the symmetry of geometric figures and vice versa. This is not particularly startling, but we have given a mathematical statement of this fact. The symmetric structure in each case (the water molecule and the rectangle, the methyl chloride molecule and the equilateral triangle) is expressed mathematically by an *abstract* group: a set of, in these cases, 4 or 6 elements which can be combined in certain ways. Whatever we can deduce about the abstract group can be applied to any situation in which it manifests itself physically.

Before we leave this section, we take the opportunity to point out an idea which we shall use in the second unit on groups. Look again at the group table for the equilateral triangle. In the top left-hand corner, we have this pattern:

$\circ$	e	R_1	R_2	S_1	S_2	S_3
e	e	R_1	R_2	—	—	—
R_1	R_1	R_2	e	—	—	—
R_2	R_2	e	R_1	—	—	—
S_1	—	—	—	—	—	—
S_2	—	—	—	—	—	—
S_3	—	—	—	—	—	—

We see that $\{e, R_1, R_2\}$, with the group operation, forms a group in its own right. It is a *subgroup* of the parent group.

Definition 2
* * *

A subgroup of a group is a subset which is itself a group (for the same binary operation as in the parent group). How do we test for subgroups? We can, of course, just check that the four group axioms are satisfied. But this is rather wasteful of effort. For example, we know in advance that the binary operation is associative.

Exercise 2
(4 minutes)

Exercise 2

Show that for any group, a subset forms a subgroup, provided that the closure and inverse axioms hold. ■

Solution 2

Solution 2

Let H be the subset of the group and $\circ$ be the binary operation.

Of the four group axioms, we are given that H is closed and that each element has an inverse, and we do not need to test for associativity. So the only work we really have to do is to show that H contains an identity. Suppose $h \in H$ and that h_1 is the inverse of h; we are given that $h_1 \in H$ and that $h \circ h_1 \in H$, since H is closed. But $h \circ h_1$ = the identity element, and so H contains the identity. ■

30.2.5 Properties of Groups

30.2.5

Main Text
* * *

It is not possible (or desirable) to attempt to develop any sort of systematic treatment of group theory in this course, but it is of interest to see just a few results to illustrate the approach and the methods.

To save you having to refer back, we list the four group axioms again.

A set G with an operation $\circ$ (denoted by $(G, \circ)$, as usual) is a group if

(i) $\circ$ is closed;
(ii) $\circ$ is associative;
(iii) $\exists_e \forall_g \, g \circ e = e \circ g = g \qquad (e \in G, g \in G)$
e is called the *identity element* of G;
(iv) given any element $g \in G$,

$$\exists_h \qquad g \circ h = e \qquad (h \in G)$$

h is called the *inverse* of g.

(The symbols $\exists_x$ and $\forall_x$ were introduced in *Unit 17*.)

We have had a little experience in this course of making deductions from a set of axioms, but we can hardly call it familiarity. This is an art which has to be learnt in mathematics. Here, we have four particularly simple axioms which have a good deal in common with the real numbers and we can often use the similarity as the intuitive basis for a formal argument. All that follows will be in the context of some given group $(G, \circ)$.

There is one point which needs tidying up first of all. In axiom (iv) we have used the identity mentioned in (iii) to define an inverse element, implying that e is unique. This is the case, and we shall now prove it as a simple example of "axiom manipulation".

Example 1

Example 1

PROPOSITION 1

In any group there is a unique identity element.

PROOF

We suppose that there are two identity elements e and f. Then if g is any element of G, we have

$$e \circ g = f \circ g = g \qquad \text{(axiom (iii))}.$$

In particular, we can take $g = f$, and then

$$e \circ f = f \circ f = f.$$

Now we have supposed that f is an identity, so by axiom (iii) again

$$e \circ f = e$$

whence

$$e = f.$$

■

This justifies our talking in future about *the* identity element of a group.

Example 2

Example 2

In ordinary arithmetic, if a is the inverse of b, then b is the inverse of a. This result can also be proved here.

PROPOSITION 2

If g_1 is the inverse of g, then g is the inverse of g_1.

PROOF

We are given that g_1 is the inverse of g, so that

$$g \circ g_1 = e \qquad \text{(axiom (iv))}$$

g_1 itself has an inverse g_2 (axiom (iv)), and

$$\begin{aligned}(g \circ g_1) \circ g_2 &= e \circ g_2 \\ &= g_2 \qquad \text{(axiom (iii))}\end{aligned}$$

Now

$$\begin{aligned}g \circ (g_1 \circ g_2) &= g \circ e \qquad \text{(axiom (iv))} \\ &= g \qquad \text{(axiom (iii))}\end{aligned}$$

But

$$(g \circ g_1) \circ g_2 = g \circ (g_1 \circ g_2) \qquad \text{(axiom (ii))}$$

so

$$g = g_2,$$

which proves that the inverse of g_1 is $g_2 = g$. ■

Discussion
★ ★

The inverse of g is usually written either as $\tilde{g}$ or g^{-1}. The notation g^{-1} has the disadvantage that it looks like a multiplicative inverse of a real number. On the other hand, it is widely used in group theory and it is a fact of mathematical life that one has to get used to different notations for the same thing *and* also the same notation for different things. (We have already had the example in matrices where we write AB as the "product" of matrices A and B, even though matrix multiplication does not have several of the properties of ordinary multiplication of numbers.) We shall use the notation g^{-1}. ■

Exercise 1

Exercise 1
(5 minutes)

Prove the following propositions:

(i) PROPOSITION 3

$$g^{-1} \circ g = e$$

(HINT: Use Proposition 2.)

(ii) PROPOSITION 4
The inverse of g is unique. ■

Solution 1

Solution 1

(i) We have shown in Proposition 2 that an inverse of g^{-1} is g, which means

$$g^{-1} \circ g = e.$$

We have

$$g \circ g^{-1} = g^{-1} \circ g = e.$$

(ii) Notice that we have usually referred to *the* inverse of g, which was premature. Our proofs are still sound, however, because we have never *used* the uniqueness of the inverse. We now prove that each element of a group has a *unique* inverse.

Let g^{-1} and $\tilde{g}$ be two inverses of g; then

$$g \circ g^{-1} = g \circ \tilde{g} = e \qquad \text{(definition of inverse)}$$

and

$$g^{-1} \circ g = \tilde{g} \circ g = e \qquad \text{(by part (i) above).}$$

Hence

$$g^{-1} \circ (g \circ g^{-1}) = g^{-1} \circ (g \circ \tilde{g})$$

i.e.

$$(g^{-1} \circ g) \circ g^{-1} = (g^{-1} \circ g) \circ \tilde{g} \qquad \text{(axiom (ii))}$$

i.e.

$$e \circ g^{-1} = e \circ \tilde{g} \qquad \text{(Proposition 3)}$$

or

$$g^{-1} = \tilde{g} \qquad \text{(axiom (iii)).}$$

■

Discussion
* *

As with all mathematics, the art of proof from axioms can only be learnt by practice, and it is unlikely that you have had much practice in the past. Even in traditional school geometry, which is the closest in method to our present subject, we have a signpost in the form of a diagram which may not only suggest what to prove, but often the method of proof.

Unfortunately, in group theory, we usually have neither a suggestion of what to prove nor of how to prove it. This is usually only due to inexperience: after a while we learn to prove (or disprove) a suggestion, and later even to make our own suggestions.

The following exercises are meant for practice.

Where you find it necessary to use our solution, then it is advisable to come back to the exercise, after a lapse of time, to see if you can do without the solution. At the very least, you should state the axioms and propositions (which we have omitted) used in each step of the solution.

Exercise 2

Exercise 2
(5 minutes)

(i) Prove that

$$(g \circ h)^{-1} = h^{-1} \circ g^{-1}.$$

(ii) Prove that the equation $g \circ x = h$, where g and h are known elements of G, has a unique solution. What is the solution? ■

Main Text
* *

Although we gave some intuitive justification for the four axioms we chose for a group, the real justification lies in the historical development of group theory.

The axioms are by no means unique. We could, for instance, replace them by the following three axioms: G is a set with a binary operation $\circ$ defined on it such that

($G1$) $\circ$ is closed;
($G2$) $\circ$ is associative;
($G3$) the equations $g \circ x = h$ and $y \circ g = h$ each have unique solutions in G for every $g, h \in G$.

The four axioms can be deduced from these three and vice versa.

The axioms $G1$, $G2$, $G3$ are said to be equivalent to our original set of axioms. There are many different equivalent axiom systems for a group. The four axioms we have chosen are probably the most convenient to work with.

When mathematicians are faced with a set of axioms, two of the questions they tend to ask are the following:

What happens if we add additional axioms?

What happens if we throw away some of the axioms?

In the first case, the most important is the axiom

$$\text{for any } g_1, g_2 \in G, g_1 \circ g_2 = g_2 \circ g_1$$

i.e. to require $\circ$ to be commutative on G. Such groups are called commutative groups or Abelian groups after the Scandinavian mathematician Niels Henrik Abel (1802–1829), who profoundly affected the direction and content of mathematics, even though he died in his twenties. Notice that not all groups are Abelian. (Note also that not all mathematicians who are interested in groups die young!)

Definition 1
* * *

Niels Henrik Abel

In the second case the axioms that "go first" are usually the two involving the identity (axioms (iii) and (iv)). We are then left with a closed associative binary operation on the set G to which is attached the name *semi-group*. Although it is a very general structure, there has been a good deal of recent activity in this subject, because of its importance, among other places, in the theory of "context free languages" which turn up in the study of computer programming languages.

Solution 2

Solution 2

(i) $(g \circ h) \circ (h^{-1} \circ g^{-1}) = ((g \circ h) \circ h^{-1}) \circ g^{-1}$

$$= (g \circ (h \circ h^{-1})) \circ g^{-1}$$

$$= (g \circ e) \circ g^{-1}$$

$$= g \circ g^{-1}$$

$$= e.$$

So $h^{-1} \circ g^{-1}$ is the inverse of $g \circ h$, and since the inverse of an element is unique,

$$(g \circ h)^{-1} = h^{-1} \circ g^{-1}.$$

(ii) Suppose that there are two solutions x_1 and x_2; then

$$g \circ x_1 = g \circ x_2 = h$$

whence

$$g^{-1} \circ (g \circ x_1) = g^{-1} \circ (g \circ x_2)$$

i.e.

$$(g^{-1} \circ g) \circ x_1 = (g^{-1} \circ g) \circ x_2$$

$$e \circ x_1 = e \circ x_2$$

$$x_1 = x_2.$$

The solution is obtained from

$$g \circ x = h$$

whence

$$g^{-1} \circ (g \circ x) = g^{-1} \circ h$$

i.e.

$$(g^{-1} \circ g) \circ x = g^{-1} \circ h$$

$$e \circ x = g^{-1} \circ h$$

so

$$x = g^{-1} \circ h$$

so the unique solution is $g^{-1} \circ h$. ■

30.2.6

Main Text

★ ★

30.2.6 Permutations and Cayley's Theorem

There are a number of elementary but far-reaching results in group theory, and in this final section of the unit we lead up to just one of these: it is known as Cayley's Theorem.

In section 30.2.2 we saw how a symmetry operation defines a mapping from the set of symmetry operations to itself. As we noted there, each symmetry operation occurs once and once only in each row of the table, so the mapping is one-one and could be regarded as a rearrangement of the set of symmetry operations. We called such a function a *permutation*.

In section 30.2.2 we used as an example the symmetry group of the triangle and saw that the element S_1 defined a permutation p, where

$$p:e \longmapsto R_1 \circ e = R_1$$

$$p:R_1 \longmapsto R_1 \circ R_1 = R_2$$

$$p:R_2 \longmapsto R_1 \circ R_2 = e$$

$$p:S_1 \longmapsto R_1 \circ S_1 = S_3$$

$$p:S_2 \longmapsto R_1 \circ S_2 = S_1$$

$$p:S_3 \longmapsto R_1 \circ S_3 = S_2$$

We can do the same sort of thing for a general group $(G, \circ)$. A given element h can be used to define a mapping p_h of G to itself, where

$$p_h:g \longmapsto h \circ g \qquad (g \in G).$$

We can use the group axioms to show that this mapping is one-one; for if p_h did map g and g_1 to the same image, then $h \circ g$ and $h \circ g_1$ would be equal, whence

$$h^{-1} \circ (h \circ g) = h^{-1} \circ (h \circ g_1)$$

whence

$$(h^{-1} \circ h) \circ g = (h^{-1} \circ h) \circ g_1$$

i.e.

$$e \circ g = e \circ g_1$$

i.e.

$$g = g_1.$$

Thus p_h maps distinct group elements to distinct group elements, and the mapping is one-one as claimed.

Furthermore, we can show that the image of G under the mapping p_h is G, since for any element $g \in G$,

$$p_h:h^{-1} \circ g \longmapsto h \circ (h^{-1} \circ g) = g.$$

Thus each h gives us a p_h which is a one-one mapping of G to itself, i.e. $p_h(G) = G$; that is, p_h is a permutation which gives a rearrangement of the set G.

Exercise 1
(3 minutes)

Exercise 1

Show that, if h and k are distinct elements of G, then the mappings p_h and p_k are distinct. ■

Solution 1

Solution 1

For two functions f and g to be equal, they must have the same domain and $f(x) = g(x)$ for *all* x in the domain. For p_h and p_k we have, for instance,

$$p_h(e) = h$$

and

$$p_k(e) = k$$

and so p_h and p_k cannot be equal. ■

Main Text
* * *

Given two permutations p_h, p_k of a group $(G, \circ)$ we can combine them by composition to obtain

$$p_h \circ p_k : g \longmapsto h \circ (k \circ g) \qquad (g \in G).$$

(We use $\circ$ to denote composition on the set of permutations, to distinguish it from the operation defined on G.) Since

$$h \circ (k \circ g) = (h \circ k) \circ g$$

we have

$$p_h \circ p_k = p_{h \circ k}.$$

So combining elements in G corresponds to combining the corresponding permutations. That is, the function

$$p : g \longmapsto p_g \qquad (g \in G)$$

is a morphism of $(G, \circ)$ to the set of permutations of G under the composition of functions.

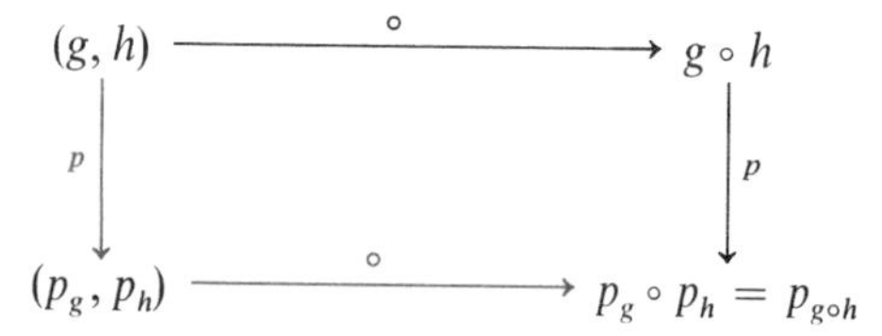

From Exercise 1 we know that different g's correspond to different p_g's, so this morphism is an isomorphism; that is, it is just a relabelling of the elements of G. So the set of all permutations associated with the elements of a group $(G, \circ)$ themselves form a group (under composition), which is isomorphic to $(G, \circ)$. We have the following theorem.

THEOREM

Theorem
* * *

Let $(G, \circ)$ be a group; then $(G, \circ)$ is isomorphic to a group of permutations on G, under composition.

Example 1

Example 1

For any *set* we can consider the set of *all* rearrangements. (We do not restrict the set to be a group.) A rearrangement is defined by a permutation, that is a one-one function which maps the set *on to itself*. (By *on to itself* we mean that the image set under the function is the domain of the function.) Two permutations can be combined by composition, and the set of all permutations is then a group. We can verify this as follows:

(i) The combination of two permutations has the effect of rearranging the set twice: the result is still a rearrangement of the original set, so the combination of two permutations is a permutation (closure).
(ii) Composition of functions is associative (associativity).
(iii) If e stands for the permutation which maps every element to itself, then

$$e \circ f = f \circ e = f$$

for every permutation f (identity).

(iv) For any permutation f, there is another permutation f^{-1} which is the inverse function of f, and

$$f \circ f^{-1} = e$$

■

(inverse).

Example 2

Example 2

Consider a set with four elements $\{A, B, C, D\}$.

The total number of rearrangements that can be made of the four letters is 24. The complete group of permutations of four objects contains 24 elements. We can spot straight away that this group must have a subgroup of six elements because we can consider all those permutations which leave D unchanged. We are then left with all the possible rearrangements of A, B and C which we know gives rise to a group of six permutations. In the same way we could regard A or B or C fixed and obtain three more subgroups of six permutations. ■

Exercise 2

Exercise 2
(5 minutes)

Consider the expression

$$x_1 x_2 x_3 + x_4,$$

where the x's stand for real numbers.

The permutation defined by

$$(1, 2, 3, 4) \longmapsto (4, 1, 2, 3)$$

i.e.

$$1 \longmapsto 4,\ 2 \longmapsto 1, \text{ etc.},$$

can be interpreted as mapping this expression to

$$x_4 x_1 x_2 + x_3,$$

which is *not* the same as the original expression. The permutation defined by

$$(1, 2, 3, 4) \longmapsto (2, 1, 3, 4)$$

maps the original expression to

$$x_2 x_1 x_3 + x_4$$

which *is* the same as the original expression. Out of the 24 rearrangements of (1, 2, 3, 4), some give us the original expression and some do not. Find the ones that do leave it unchanged. Show that they form a group; write down its group table.

The group is called the *symmetry group* of the expression. Symmetry is used here in the same sense as in the geometrical usage: the expression is mapped to itself by a permutation from the symmetry group. ■

Main Text
* * *

In order to see why this permutation idea is important and useful we make the following observations. Assume that the group G we are considering contains a finite number of elements, n say, and write the elements of G in a row thus:

$$(g_1, g_2, \ldots, g_n).$$

If we perform the permutation p_g where $g \in G$, then the images under this mapping are

$$(g \circ g_1, g \circ g_2, \ldots, g \circ g_n),$$

which is some rearrangement of the elements $g_1, \ldots, g_n$. *But* every rearrangement of the list $(g_1, \ldots, g_n)$, not only those corresponding to

(*continued on page 47*)

Solution 2

Solution 2

The expression is left unchanged by the following permutations.

$$e:(1,2,3,4) \longmapsto (1,2,3,4)$$

$$a:(1,2,3,4) \longmapsto (2,3,1,4)$$

$$b:(1,2,3,4) \longmapsto (3,1,2,4)$$

$$c:(1,2,3,4) \longmapsto (1,3,2,4)$$

$$d:(1,2,3,4) \longmapsto (2,1,3,4)$$

$$f:(1,2,3,4) \longmapsto (3,2,1,4).$$

Note that, for example, performing the permutation c means

> leave the elements in the 1st and 4th positions unchanged; interchange the elements in the 2nd and 3rd positions.

Hence

$$c:(2,3,1,4) \longmapsto (2,1,3,4),$$

so

$$c \circ a:(1,2,3,4) \longmapsto (2,1,3,4),$$

i.e.

$$c \circ a = d.$$

The group table is

	e	a	b	c	d	f
e	e	a	b	c	d	f
a	a	b	e	f	c	d
b	b	e	a	d	f	c
c	c	d	f	e	a	b
d	d	f	c	b	e	a
f	f	c	d	a	b	e

(Remember that $a \circ b$ means carry out the permutation b first and then a, and the entry $a \circ b$ goes in row a, column b.)

This table is in fact a relabelled version of the table we had for the symmetry operations of the equilateral triangle. ■

(*continued from page 45*)

mappings such as p_g, can be effected by thinking of the list as a vector, and multiplying by a *permutation* matrix which has only 0's and 1's as elements, and only one 1 in each column and row. For example:

$$(g_1, g_2, g_3)\begin{pmatrix} 0 & 1 & 0 \\ 0 & 0 & 1 \\ 1 & 0 & 0 \end{pmatrix} = (g_3, g_1, g_2).$$

It is not hard to believe (we offer no proof) that every permutation of n symbols can be thought of as an $n \times n$ permutation matrix of this form, and that these permutation matrices form a group under multiplication. Stretching your faith just a little further, you can see that *every group* can be looked upon as a group of permutation matrices, and it is just this idea which leads to the important area of *group representation theory*, which uses matrix methods to tackle these abstract groups.

Combining the result of Example 1 with the theorem on page 44, we obtain

CAYLEY'S THEOREM

Cayley's Theorem
* * *

Any finite group $(G, \circ)$ of order n is isomorphic to a subgroup of the group of all permutations on a set of n elements.

We have in fact proved a more general theorem than Cayley's: there is no need to restrict the group to be finite. We have the following theorem:

THEOREM

Theorem
* * *

Any group $(G, \circ)$ is isomorphic to a subgroup of the group of all permutations on the set G.

The importance of this theorem is its universality. Every group can be regarded as a permutation group: if we study permutation groups we are studying group theory. That is, by studying the particular (permutation groups) we are studying the general.

Arthur Cayley

Looking back on our journey, we see that we have come a long way: we started with simple ideas of symmetry in the drawing of graphs and in the world around us; symmetry was then observed, analysed and abstracted until we obtained the four group axioms. We played a little with the axioms and proved one outstanding result, Cayley's Theorem.

You may have found the abstract mathematics difficult. This seems to be the major difficulty with this sort of algebra. The algebra involves a great deal of thought, but hardly any manipulative ability. A proof may be a couple of trivial lines, but the organisation of those two lines is extraordinarily difficult at the beginning. Algebra, released from being just symbolic arithmetic, is a new subject with new patterns of thought, which it takes time to acquire.

Unit No.		Title of Text
1		Functions
2		Errors and Accuracy
3		Operations and Morphisms
4		Finite Differences
5	NO TEXT	
6		Inequalities
7		Sequences and Limits I
8		Computing I
9		Integration I
10	NO TEXT	
11		Logic I — Boolean Algebra
12		Differentiation I
13		Integration II
14		Sequences and Limits II
15		Differentiation II
16		Probability and Statistics I
17		Logic II — Proof
18		Probability and Statistics II
19		Relations
20		Computing II
21		Probability and Statistics III
22		Linear Algebra I
23		Linear Algebra II
24		Differential Equations I
25	NO TEXT	
26		Linear Algebra III
27		Complex Numbers I
28		Linear Algebra IV
29		Complex Numbers II
30		Groups I
31		Differential Equations II
32	NO TEXT	
33		Groups II
34		Number Systems
35		Topology
36		Mathematical Structures

Cut-out Diagrams

To make the Platonic solids, cut out the following diagrams, and fold along all the interior edges. Now apply cellotape!

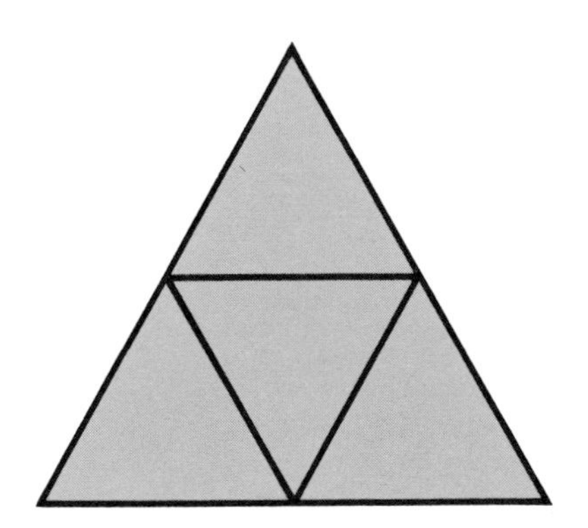

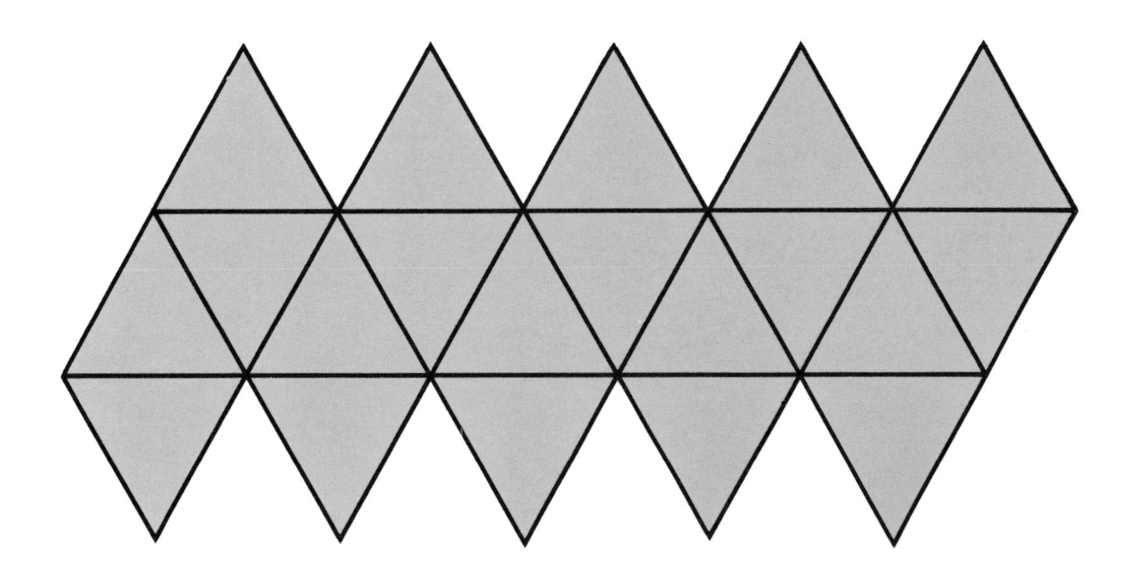

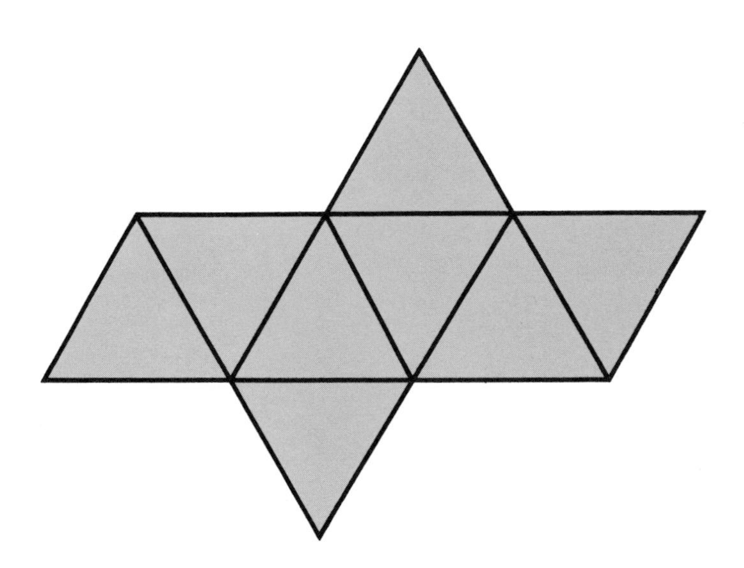

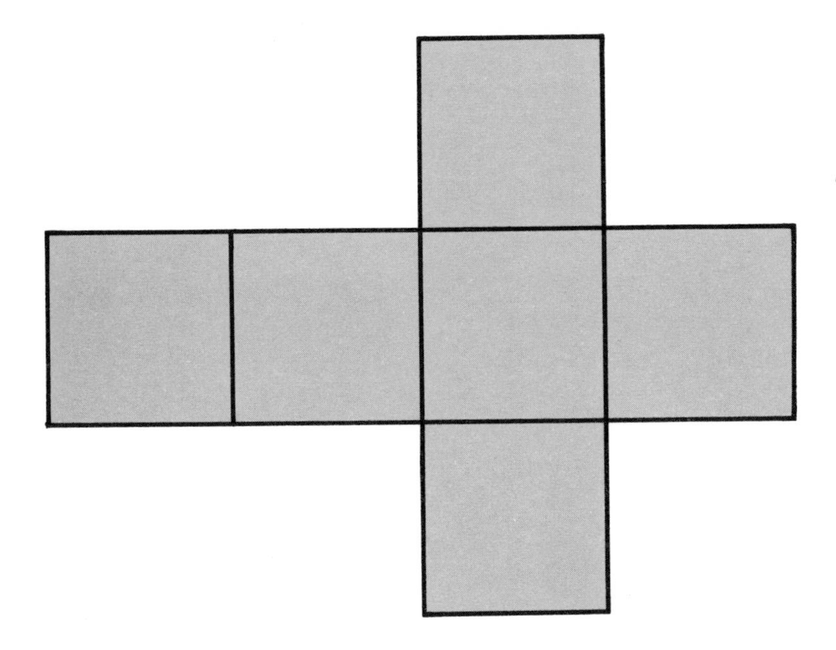

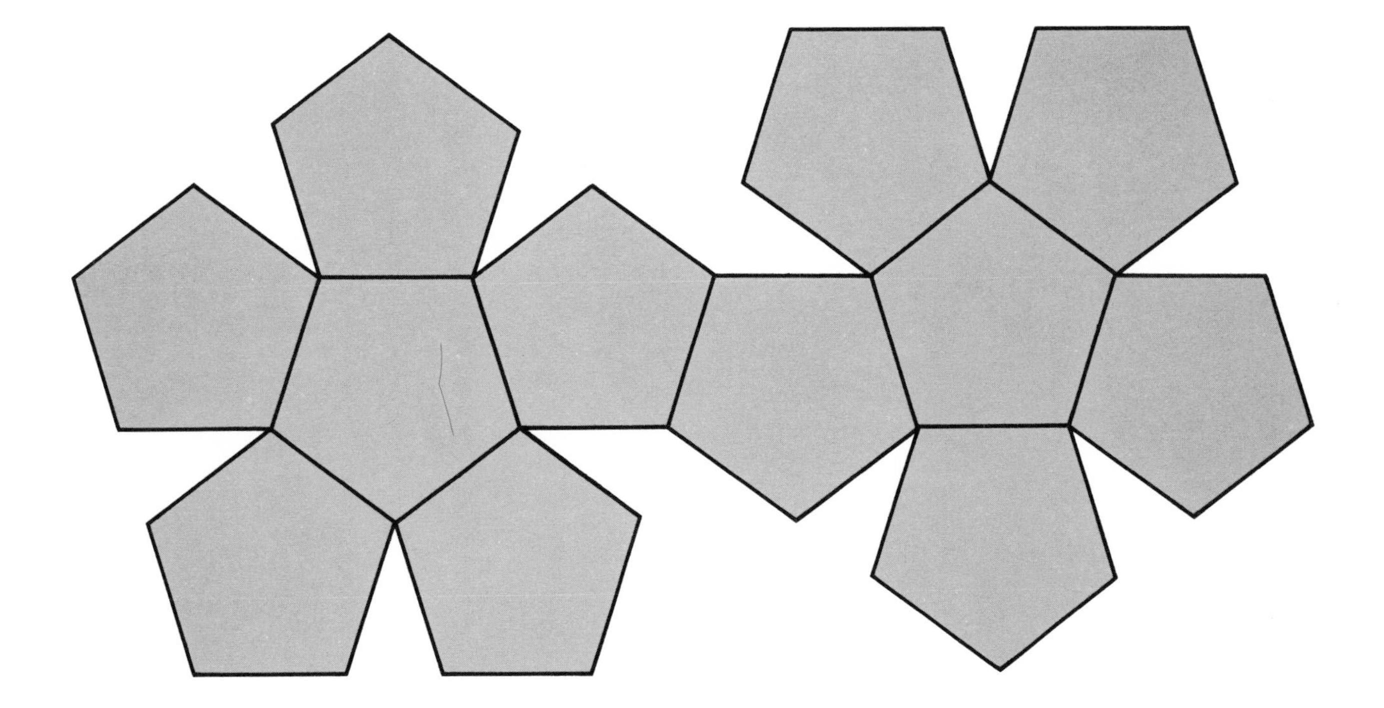